T0104667

# GENE EDITING, EPIGENETIC, CLONING AND THERAPY

Amin Elser

authorHOUSE®

*AuthorHouse™*
*1663 Liberty Drive*
*Bloomington, IN 47403*
*www.authorhouse.com*
*Phone: 1 (800) 839-8640*



*Published by AuthorHouse 08/04/2016*

*ISBN: 978-1-5246-2199-5 (sc)*
*ISBN: 978-1-5246-2198-8 (e)*

*Print information available on the last page.*



*This book is printed on acid-free paper.*

# Contents

# Introduction

This book is really helpful for someone who wants to start learning about genes and DNA. It is well written book describing all the introductory materials one would need to become current with genomes and genomics topics. It begins with an introduction to DNA and genes in Chapter 1, and goes on from there through epigenetic in chapter 2, including acetylation, methylation, ubiquitylation of protein, deimination and proline isomerization. It goes through gene editing in chapter 3 which includes good description of TALENs, ZFNs and CRISPR/Cas systems. Chapter 4 includes cloning using artificial embryo twinning, somatic cell nuclear transfer, and asexual reproduction. Chapter 5 is the material on basic stem cells of embryonic stem cells and adult stem cells. Chapter 6 discusses techniques and technology of gene therapy and cloning therapy. Chapter 7 includes descriptions on cell division, mitosis, meiosis, biological life cycle, parthenogenesis, bacterial conjugation, DNA fingerprints, genetic relationship between individuals and surname studies.

The book includes many diagrams and glossary and an index at the front. For a serious book on DNA and genes this book is quite readable it is a user-friendly textbook, so that many readers will find it helpful to read some chapters more than once. The book is a valuable introduction to the extremely important field of genes and genomics.

A highly informative book, I recommend it highly to anyone interested in the subject."

# CHAPTER 1

# DNA and GENES

## 1.1 Chromosomes

In sexually reproducing organism two types of cell division are needed. One is for the processes of growth, repair and asexual reproduction and it is called mitosis. Mitosis produces daughter cells that are diploid and genetically identical to the parent cell.

When the organism wants to make gametes (eggs and sperms) a different mechanism is required. Gametes are not diploid like all the other body cells, but instead they only have one member of each homologous pair of chromosomes. In order to make a haploid daughter cell, a second type of cell division, meiosis, is needed, Figure (1).

Figure (1): Mitosis and meiosis

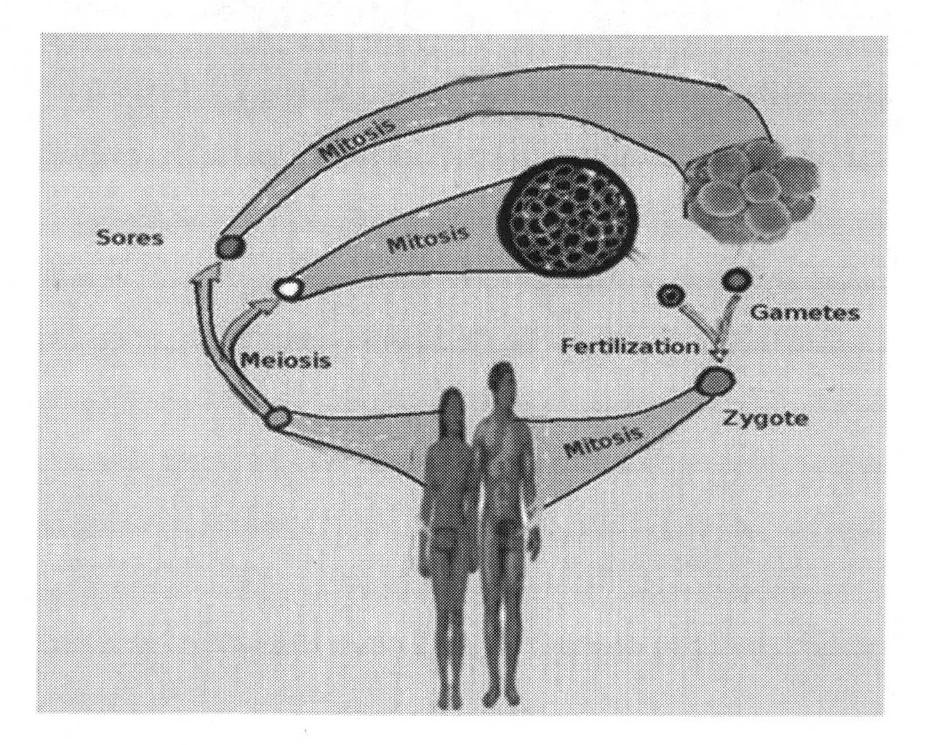

Chromosomes are thread-like structures located inside the nucleus of a cell.. Each chromosome is made of protein and a single molecule of deoxyribonucleic acid (DNA). Passed from parents to offspring, DNA contains the specific instructions that make each type of living organism unique.

The nucleus of each cell in our bodies contains approximately 1.8 meters of DNA in total, although each strand is less than one millionth of a centimeter thick. This DNA is tightly packed into structures called chromosomes, which consist of long chains of DNA and associated proteins. In eukaryotes, DNA molecules are tightly wound around proteins – called histone proteins - which provide structural support and play a role in controlling the activities of the genes. A strand 150 to 200 nucleotides long is wrapped twice around a core of eight histone proteins to form a structure called a nucleosome, Figure (2).

Figure (2): Nucleosome, hisone and DNA

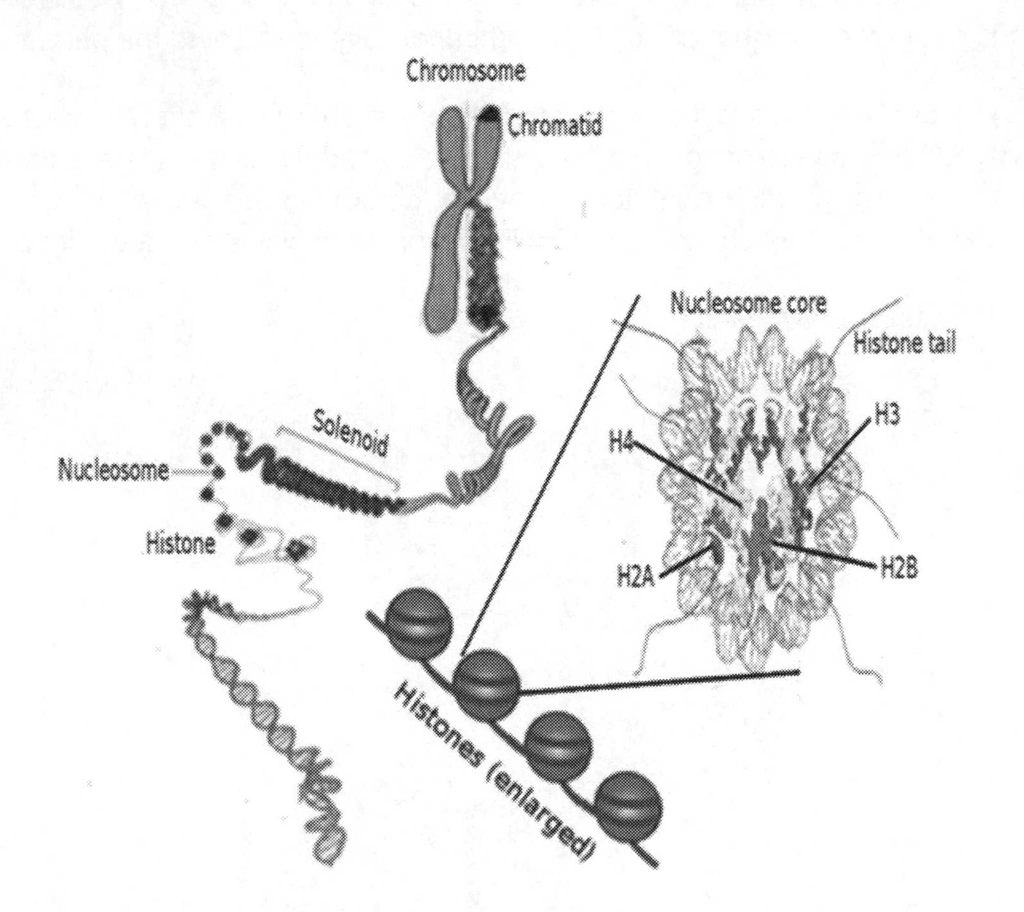


The chromosomes - and the DNA they contain - are copied as part of the cell cycle, and passed to daughter cells through the processes of mitosis and meiosis.

Chromosomes end with two telomeres. If telomeres are shortened, cells and human will age. If telomerase activity is high, or if the telomeres are extended, chromosome (and DNA) is maintained, and cellular senescence is delayed. In contrast, if telomeres are damaged or defected, cancer and certain inherited disease may be developed. Scientists found that when a cell is about to divide, the DNA molecules, which contain the four bases that form the genetic code, are copied, base by base, by DNA polymerase enzymes. However, for one of the two DNA strands, a problem exists in that the very end of the strand cannot be copied. Therefore, the chromosomes should be shortened every time a cell divides, Figure (3).

Figure (3): Shortened DNA and thus chromosome due to damaged or shortened telomere

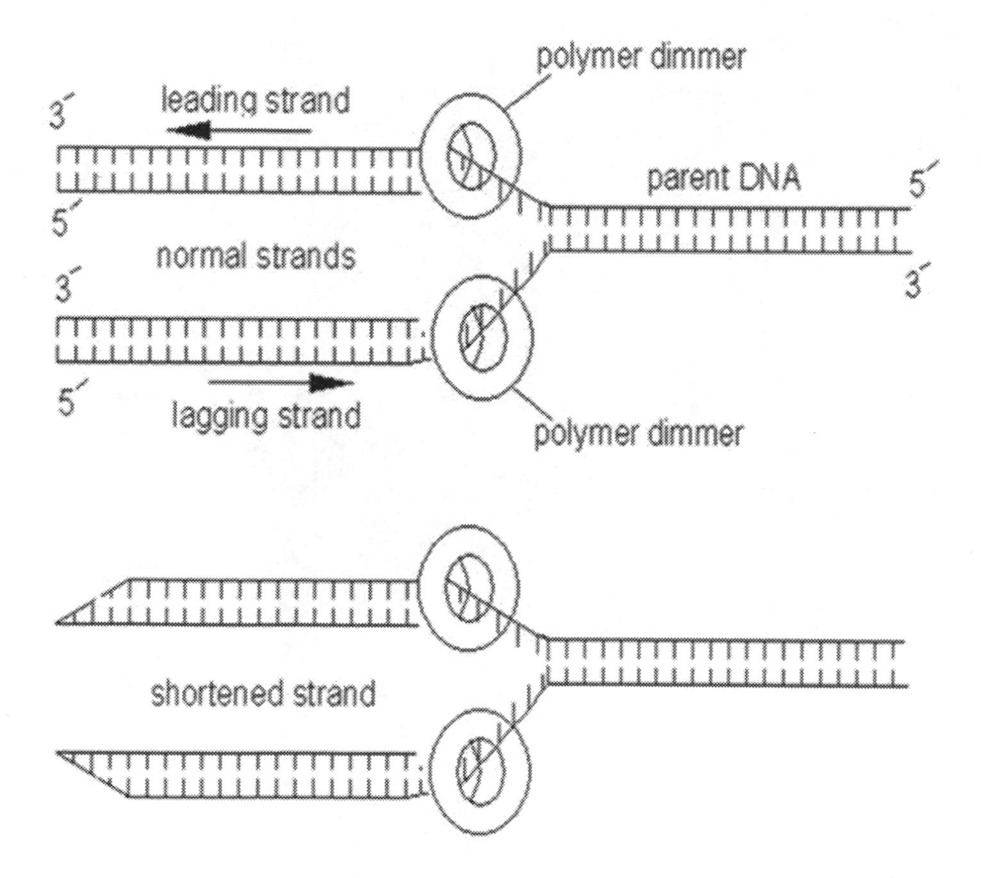


Luckily, these problems were solved when this year's Nobel Laureates (Elizabeth Blackburn, Jack Szostak, and Carol Greider) discovered how the telomere functions and found the enzyme that copies it.

During his experiment, Jack Szostak observed that a linear DNA molecule, a type of minichromosome, is rapidly degraded when introduced into yeast cells. In fact, all DNA molecules are degraded by time after replication, but putting the DNA molecules in yeast cells, they delay faster. The question is how to delay or prevent the degradation. If this is achieved, the span of life would be elongated and the aging process would be delayed.

Elizabeth Blackburn mapped DNA sequences. When studying the chromosomes of Tetrahymena, a unicellular ciliate organism, she identified a DNA sequence that was repeated several times at the ends of the chromosomes. The function of this sequence, CCCCAA (C for cytosine and A for adenine, both are nucleotides of the DNA), was unclear. Blackburn and Szostak decided to perform an experiment that would implement their discoveries jointly. From the DNA of *Tetrahymena*, Blackburn isolated the CCCCAA sequence. Szostak coupled it to the minichromosomes and put them back into yeast cells. Results were amazing - the telomere DNA sequence protected the minichromosomes from degradation, Figure (4). This was the start of the reversal of aging.

Figure (4): Szostack and Blackburn joint experiment

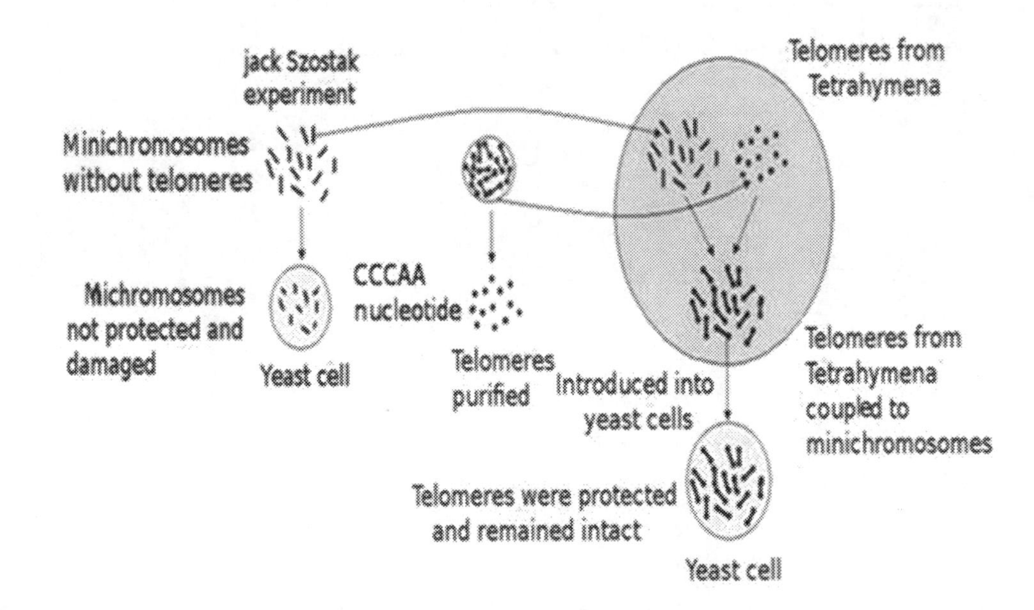


Blackburn and her graduate student Carol Greider started to investigate if the formation of telomere DNA could be due to an unknown enzyme. Greider discovered signs of enzymatic activity in a cell extract. Greider and Blackburn named the enzyme telomerase. The telomerase consists of RNA nucleotides CCCCAA and protein. Telomerase extends telomere DNA, providing a platform that enables DNA polymerases to copy the entire length of the chromosome without missing the very the very end portion, Figure (5).

Figure (5): Effect of telomerase on cell division

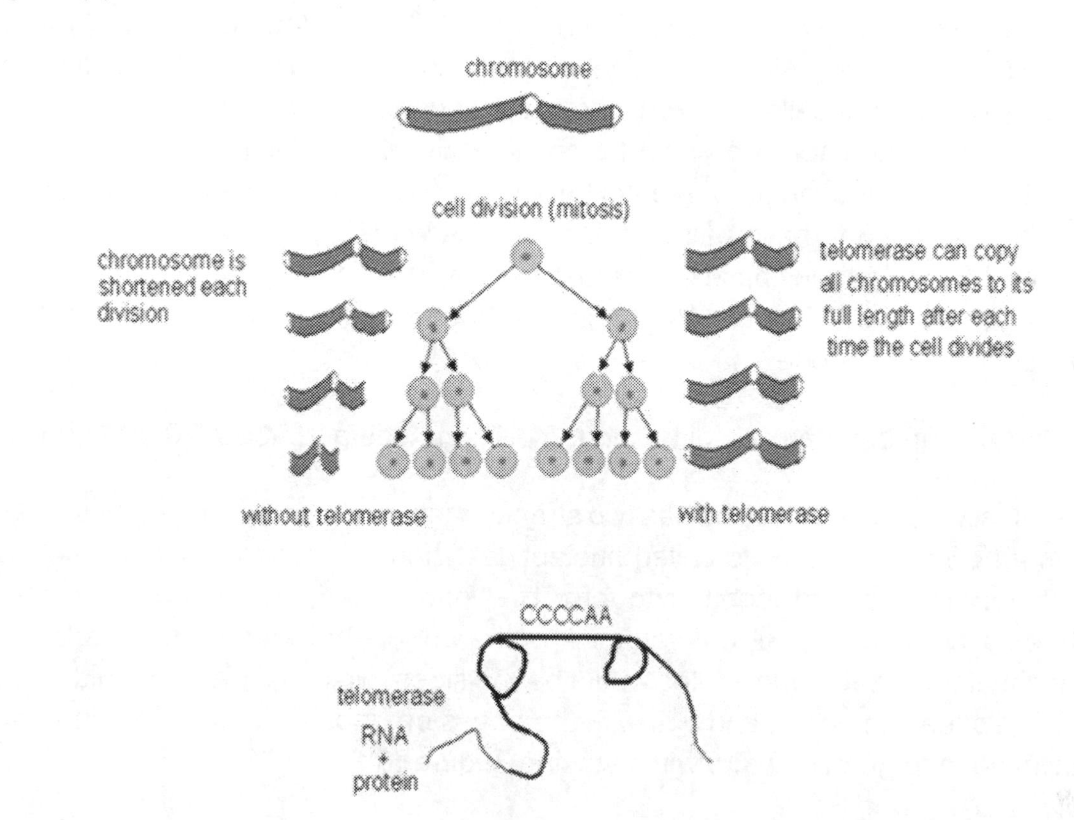


Telomeres safeguard the chromosome ends from DNA repair and degradation activities. With gradual shortening of the telomeres and chromosomes cells grew poorly and eventually stopped dividing. This could lead to premature cellular ageing – senescence. Scientists concluded that functional telomeres could prevent chromosomal damage and delay cellular senescence. They showed that the senescence of human cells is also delayed by telomerase.

### 1.1.1 DNA and RNA

Deoxyribonucleic acid or DNA is a part of chromosomes and includes information needed to develop all features of life. The DNA is found in every cell and passed down from generation to generation. DNA carries the code to control the characteristic in all life forms.

Deoxyribonucleic acid or DNA is a part of chromosomes and includes information needed to develop all features of life. The DNA is found in every cell and passed down from generation to generation. DNA carries the code to control the characteristic in all life forms.

DNA is a very long macromolcule that is the main component of chromosomes made up of molecules called nucleotides**.** Each nucleotide is composed of a nitrogen-containing nucleobase - either cytosine (C), guanine (G), adenine (A), or thymine (T)—as well as a sugar called deoxyribose and a phosphorate group. The nucleotides are joined to one another in a chain by covalent bonds between the sugar of one nucleotide and the phosphate of the next, resulting in an alternating sugar-phospate backbone. According to base pairing rules rules (A with T, and C with G), hydrogen bonds bind the nitrogenous bases of the two separate polynucleotide strands to make double-stranded DNA. Each DNA sequence that contains instructions to make a protein is known as a gene.

The entire human genome contains about 3 billion bases and about 20,000 genes.

DNA is usually a double-helix and has two strands running in opposite directions.. Each chain is a polymer of subunits called nucleotides (hence the name polynucleotide). Nucleotides are attached together to form two long strands that spiral to create a structure called a double helix. If you think of the double helix structure as a ladder, the phosphate and sugar molecules would be the sides, while the bases would be the rungs. The bases on one strand pair with the bases on another strand: adenine pairs with thymine, and guanine pairs with cytosine, Figure (6).

Figure (6): Structure of DNA

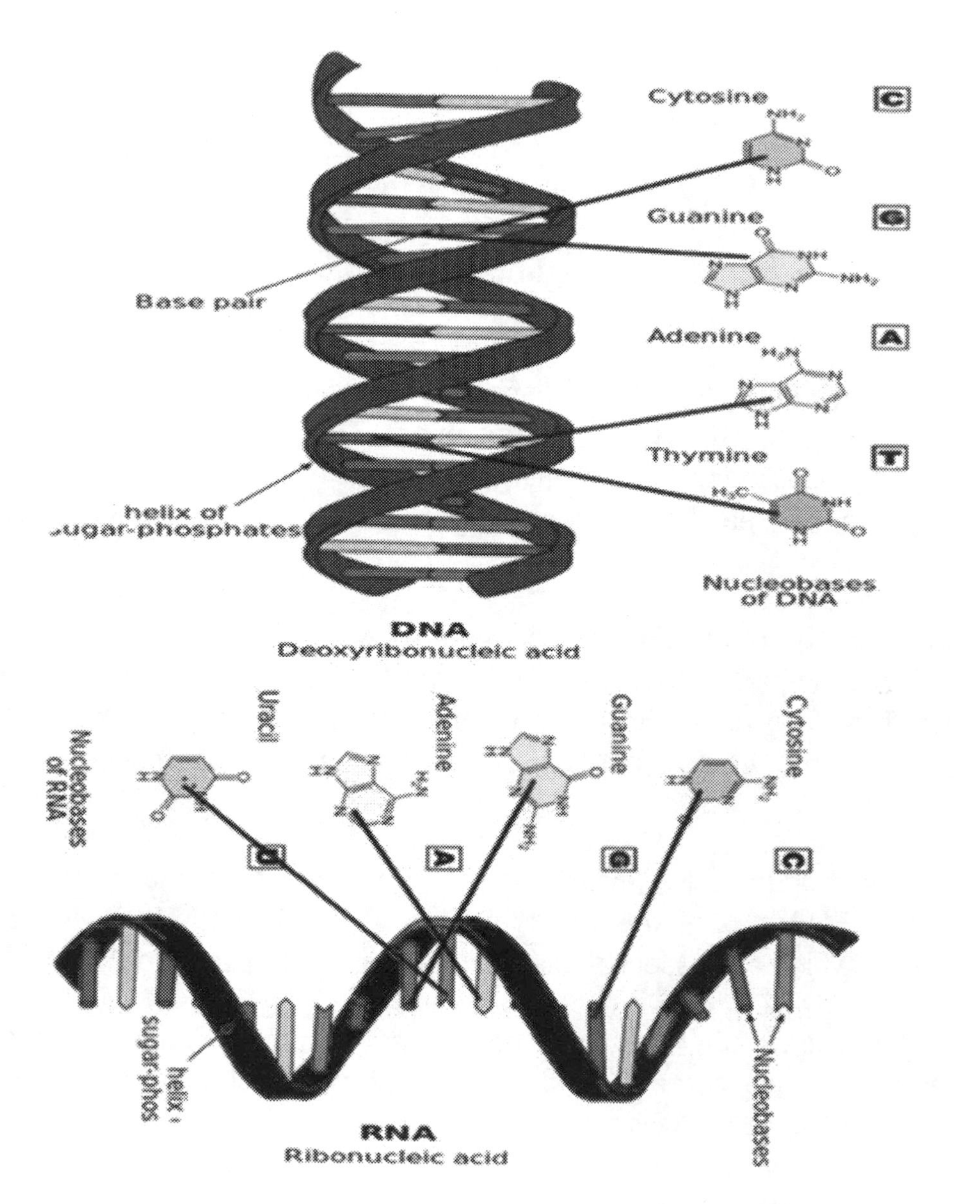


Although most DNA is packaged in chromosomes within the nucleus, mitochondria also have a small amount of their own DNA. This genetic material is known as mitochondrial DNA or mtDNA. Mitochondrial DNA (mtDNA or mDNA) is the DNA located in mitochondria, cellular organelles within eukaryotic cells that convert chemical energy from food into a form that cells can use.

The number of mitochondria in a cell can vary widely by organism, tissue, and cell type. For instance, red blood cells have no mitochondria, whereas liver cells can have more than 2000

Mitochondria have been described as “the powerhouse of the cell” because they generate most of the cell’s supply of adenosine triphosphate (ATP), used as a source of chemical energy. In addition to supplying cellular energy, mitochondria are involved in other tasks, such as signaling, cellular differentiation, and cell death, as well as maintaining control of the cell cycle and cell growth.. Each cell contains hundreds to thousands of mitochondria, which are located in the fluid that surrounds the nucleus (the cytoplasm). The number of mitochondria in a cell can vary widely by organism, tissue, and cell type. For instance, red blood cells have no mitochondria, whereas liver cells can have more than 2000.

In addition to producing energy, mitochondria store calcium for cell signaling activities, generate heat, and mediate cell growth and death. The number of mitochondria per cell varies widely; for example, in humans, erythrocytes (red blood cells) do not contain any mitochondria, whereas liver cells and muscle cells may contain hundreds or even thousands.

In addition to energy production, mitochondria play a role in several other cellular activities. For example, mitochondria help regulate the self-destruction of cells (apoptosis). They are also necessary for the production of substances such as cholesterol and heme (a component of hemoglobin, the molecule that carries oxygen in the blood).

The human mitochondrial DNA (mtDNA) is a double-stranded, and contains 37 genes coding for two rRNAs, 22 tRNAs and 13 polypeptides. Mitochondrial DNA contains 37 genes, all of which are essential for normal mitochondrial function.

Mitochondria are unusual organelles. They act as the power plants of the cell, are surrounded by two membranes, and have their own genome. They also divide independently of the cell in which they reside, meaning mitochondrial replication is not coupled to cell division. Some of these features are holdovers from the ancient ancestors of mitochondria, which were likely free-living prokaryotes.

### 1.1.2 Genes and Genomes

Proteins are assembled from amino acids using information encoded in genes. Each protein has its own unique amino acid sequence that is specified by the nucleotide sequence of the gene encoding this protein. A gene is the basic physical and functional unit of heredity. Large-scale structural variations are differences in the genome among people that range from a few thousand to a few million DNA bases; some are gains or losses of stretches of genome sequence and others appear as re-arrangements of stretches of sequence. These variations include differences in the number of copies individuals have of a particular gene, deletions, translocations and inversions. Genes, which are made up of DNA, act as instructions to make molecules called proteins. In humans, genes vary in size from a few hundred DNA bases to more than 2 million bases. The Human Genome Project has estimated that humans have between 20,000 and 25,000 genes, Figure (7).

Every person has two copies of each gene, one inherited from each parent. Most genes are the same in all people, but a small number of genes (less than 1 percent of the total) are slightly different between people

Figure (7): Genes, DNA and chromosome

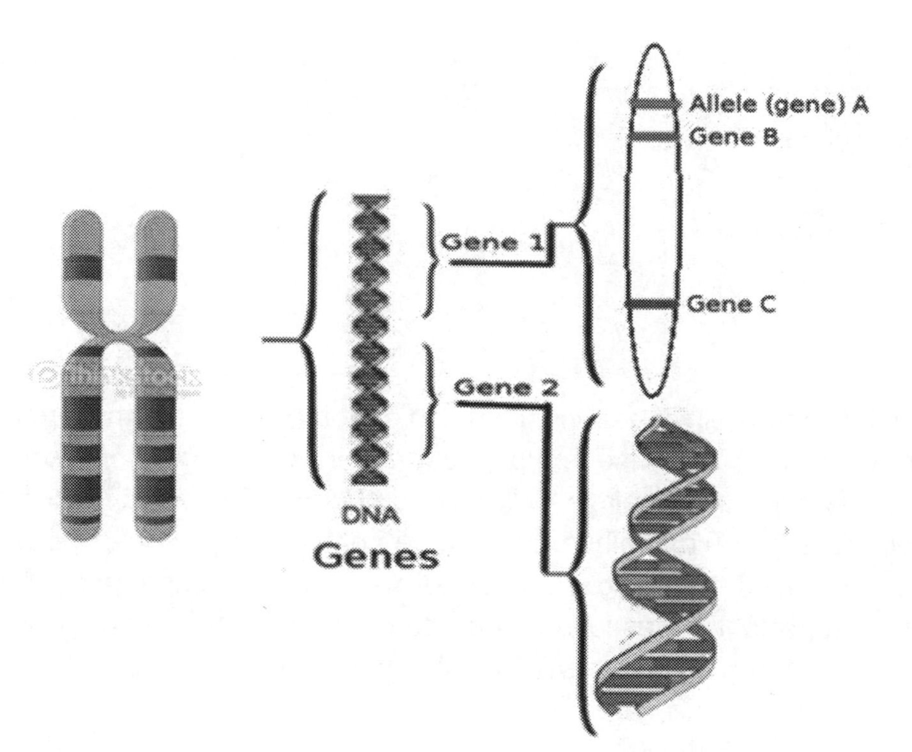


Genetic Code is the precise number and arrangement of the base pairs along the DNA that forms the organism's genetic code. Proteins are made up of amino acids which are made up of a set of nucleotide called triplets of codons. There are many combinations of amino acids that make up the proteins that are an organism's body, Figure (8).

Figure (8): Codons and amino acids

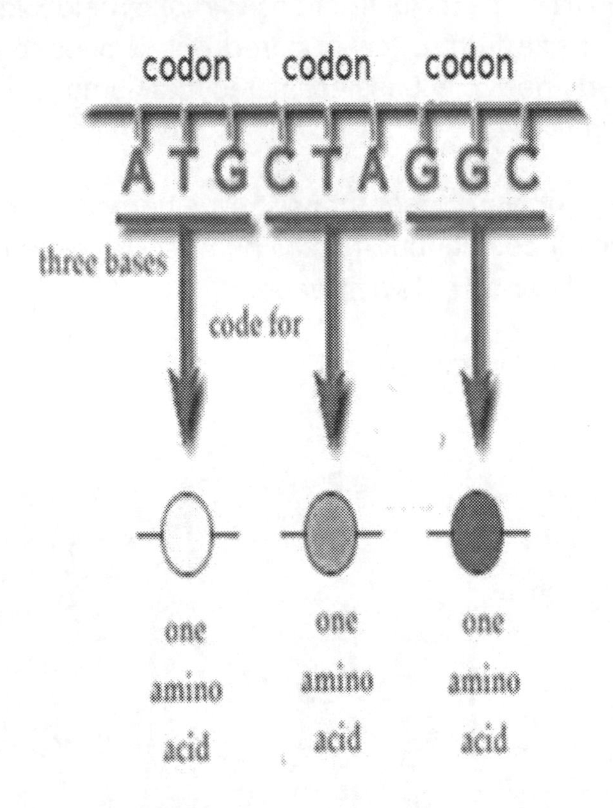


The entire DNA in the cell makes up the human genome. The Human Genome Project is an international scientific research project project with the goal of determining the sequence of chemical base pairs which make up human DNA, and of identifying and mapping all of the genes of the human genome from both a physical and a functional standpoint. It remains the world's largest collaborative biological project There are about 20,000 important genes located on one of the 23 chromosome pairs found in the nucleus or on long strands of DNA located in the mitochondria.

The DNA in the genes make up only around 2% of the genome. For some years now each of the sequences and genes discovered are carefully recorded as to their specific location, sequences etc. The whole information is stored in a database that is publicly accessible.

Nearly 13000 genes have been mapped to specific locations (loci) on each of the chromosomes.

In genomics and related disciplines, noncoding DNA sequences are components of an organism's DNA that do not encode protein sequences. Some noncoding DNA is transcriped into functional non-coding RNA molecules (e.g. transfer RNA(tRNA) and ribosome RNA (rRNA)).

The amount of noncoding DNA varies greatly among species. Where only a small percentage of the genome is responsible for coding proteins, the percentage of the genome performing regulatory functions is growing. When there is much non-coding DNA, a large proportion appears to have no biological function for the organism.

In genomics and related disciplines, noncoding DNA sequences are components of an organism's DNA that do not encode protein sequences. Such noncoding sequences are termed Junk> Some noncoding DNA is transcribed into functional non-coding RNA molecules. Other functions of noncoding DNA include the transcriptional and translational regulation of protein-coding sequences, scaffold attachment regions, origins of DNA replication, centromeres and telomeres of chromosomes.

About 1,000 base pairs would be enough DNA to encode most proteins. But introns—"extra" or "nonsense" sequences inside genes—make many genes longer than that. Human genes are commonly around 25,000 base pairs long, and some are up to 2 million base pairs.

Very simple organisms tend to have relatively small genomes. The smallest genomes, belonging to primitive, single-celled organisms, contain just over half a million base pairs of DNA.

Despite their strands, long having been considered non-coding regions of eukaryotic genomes play crucial roles in the regulation of gene expression. Non-coding regions (introns) make many genes longer than that. Human genes are commonly around 25,000 base pairs long, and some are up to 2 million base pairs. Very simple organisms tend to have relatively small genomes. The smallest genomes, belonging to primitive, single-celled organisms, contain just over half a million base pairs of DNA. A newt genome has about 15 billion base pairs of DNA, and a lily genome has almost 100 billion.

### 1.1.3 Alleles

An allele is an alternative form of a gene (one member of a pair) that is located at a specific position on a specific chromosome. Sometimes, different alleles can result in different observable phenotypic traits, such as different pigmentation. However, most genetic variations result in minute or no observable variation. These DNA codlings determine distinctive traits that can be passed on from parents to children through sexual reproduction. The process by which alleles are transmitted was discovered by Gregor Mendel.

## Dominant and Recessive Alleles

Where the heterozygote is indistinguishable from one of the homozygotes, the allele involved is said to be dominant to the other, which is said to be recessive to the former. Examples: Organisms typically have two alleles for a trait. When the alleles of a pair are heterozygous, the phenotype of one trait may be dominant and the other recessive. A population or species of organisms typically includes multiple alleles at each locus among various individuals. Allelic variation at a locus is measurable as the number of alleles (polymorphism) present, or the proportion of heterozygotes in the population. The dominant allele is expressed and the recessive allele is masked. This is known as complete dominance. In heterozygous relationships where neither allele is dominant but both are completely expressed, the alleles are considered to be co-dominant. Co-dominance is shown in AB blood type inheritance. When one allele in not completely dominant over the other, the alleles are said to express incomplete dominance.

While most genes exist in two allele forms, some have multiple alleles for a trait. A general example of this in humans is ABO blood type. Human blood type is established by the presence or absence of certain identifiers, called antigens, on the surface of red blood cells. Individuals with blood type A have A antigens on blood cell surfaces, those with type B have B antigens, where as those with type O are without antigens. ABO blood types exist as three alleles, which are signified as ($I^A$, $I^B$, $I^O$). These multiple alleles are passed from parent to offspring such that one allele is inherited from each parent.

In humans the sex chromosomes comprise one pair of the total of 23 pairs of chromosomes. The other 22 pairs of chromosomes are called autosomes, with the same gene in both members of a given pair and one pair of sex chromosomes, which are designated XX in females and XY in males. The X and Y chromosomes are physically different from one another in that the Y chromosome is much shorter, and the Y chromosome only has about nine gene loci that match those on the X chromosome. A pair of chromosomes carries the same genes in the same place, on each chromosome within the pair. However, there are different versions of a gene called alleles. These alleles may be the same (homozygous) on each pair of chromosomes, or different (heterozygous), for example, to give blue eyes or brown eyes. The figure below schematically depicts a pair of chromosomes and shows three hypothetical genes: hair color, body height, and multiple lipoma formation, Figure (9).

Figure (9): Alleles

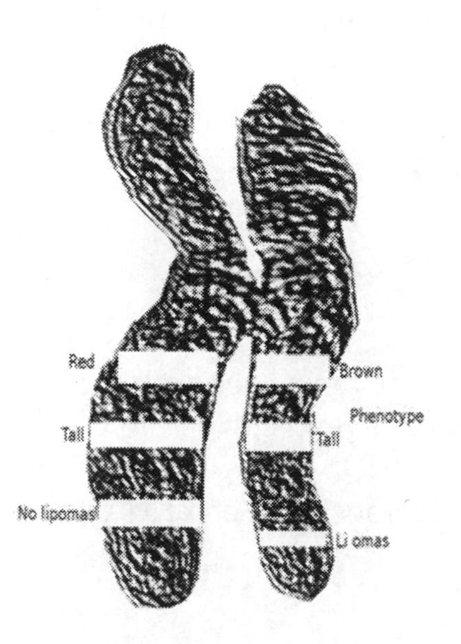


## 1.2 Protein Synthesis

Within the cell nucleus are a number of chromosomes. Each of these chromosomes contains a number of tightly coiled DNA molecules. There are numerous genes or sections strung along the DNA molecules. There are numerous genes or sections strung along the DNA molecules. A single gene provides a chemical code for the synthesis of a particular protein. Part of the DNA double helix unwinds, exposing a group of *DNA codons*. These codons consist of sets of three chemical bases.

Steps:

During transcription, a copy of the exposed DNA bases is made. A messenger RNA (mRNA) molecule then results. The mRNA molecule moves out of the nucleus, and onto the surface of a ribosome. A series of individual transfer RNA (tRNA) molecules, each attached to a certain amino acid, also move towards the ribosomes.

The tRNA molecules match their bases up against complementary bases of the mRNA molecule. The amino acids attached at the other end of the tRNAs link together via peptide bonds. The end result is a finished protein or polypeptide – a combination of "poly" amino acids connected by peptide bonds in a coded order.

Each completed protein (polypeptide) detaches from a ribosome and begins to perform its special function within the cell.

In brief, Amino acids are brought to the ribosome by tRNA. There are 20 different tRNA molecules, one for each type of amino acids. tRNA anticodons find their compliment codons on the mRNA.

mRNA codons UAA CGA GGC

tRNA anticodons AUU GCU CCG

Peptide bonds form between the amino acid forming a polypeptide. Translation stops when a stop codon is reached, Figure (10).

Figure (10): Transcription and translation of mRNA and tRNA

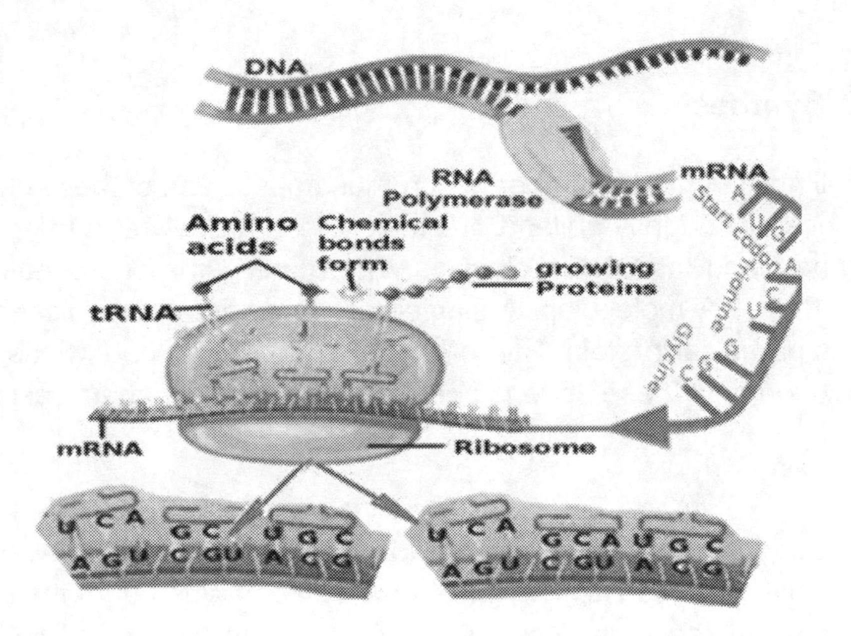


Ribosomes are the protein builders or the protein synthesizers of the cell. They are like construction guys who connect one amino acid at a time and build long chains.

## 1.3 Coding

Genes are made from a long molecule called DNA, which is copied and inherited across generations. DNA is made of simple units that line up in a particular order within this large molecule. The order of these units carries genetic information, similar to how the order of letters on a page carries information.

These two strands in the gene are labeled as template strand (promoter sequence) and coding strand. The template strand act as a blueprint for the production of the RNA transcript sequence while the coding strand is the DNA version of the transcript sequence

Transcription is the first step of gene expression, in which a particular segment of DNA is copied into RNA (mRNA) by the enzyme RNA polymerase..

Both RNA and DNA are nucleic acids, which use base pairs of nucleotides. The two can be converted back and forth from DNA to RNA by the action of the correct enzymes. During transcription, a DNA sequence is read by an RNA polymerase, which produces a complementary, antiparallel RNA strand called a primary transcript, Figure (11).

Figure (11): Coding of DNA and RNA

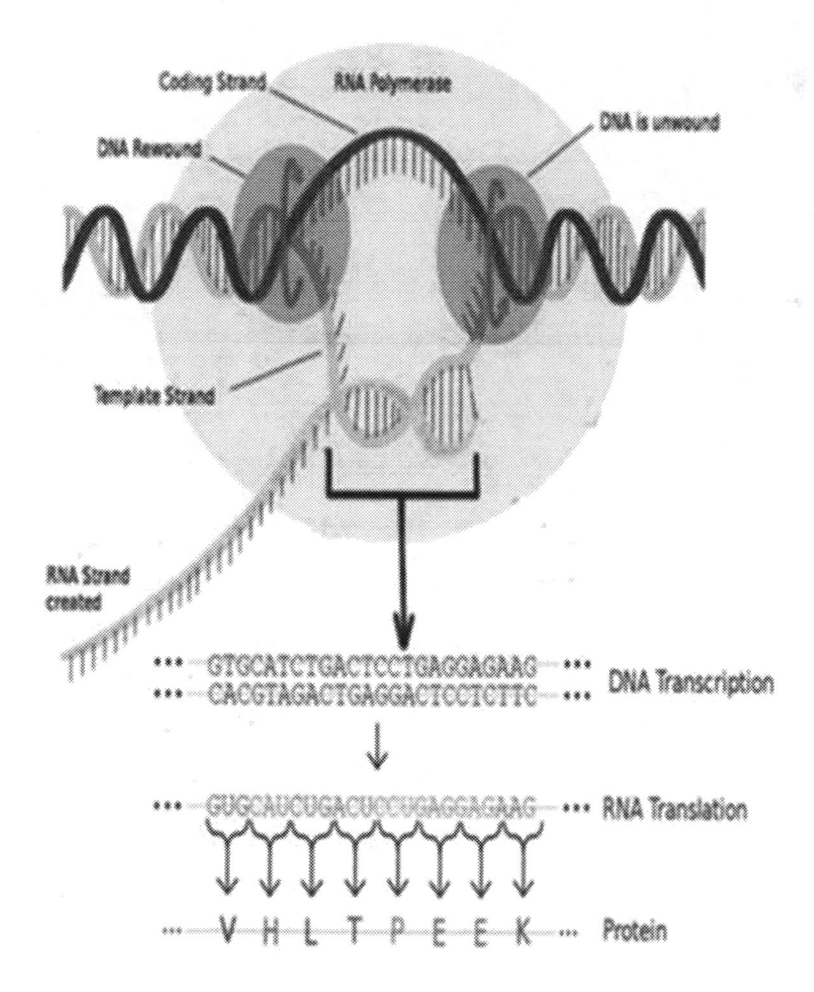

## 1.4 Control Gene Expression

The method of using the information contained in genetic material (DNA and RNA) to form protein is called gene expression.

In eukaryotes (an organism consisting of a cell or cells in which the genetic material is DNA in the form of chromosomes contained within a distinct nucleus. Eukaryotes which include all living organisms other than the eubacteria and archaebacteria) transcription takes place within the membrane-bound nucleus, and the initial transcript is adjusted before it is transported from the nucleus to the cytoplasm for translation at the ribosome s. The initial transcript in eukaryotes has coding segments (exons) alternating with non-coding segments (introns). Before the mRNA leaves the nucleus, the introns are removed from the transcript by a process called RNA splicing and extra nucleotides are attached to the ends of the transcript; these non-coding "caps" and "tails" protect the mRNA from attack by cellular enzymes and help in recognition by the ribosomes, Figure (12).

Figure (12): Gene expression

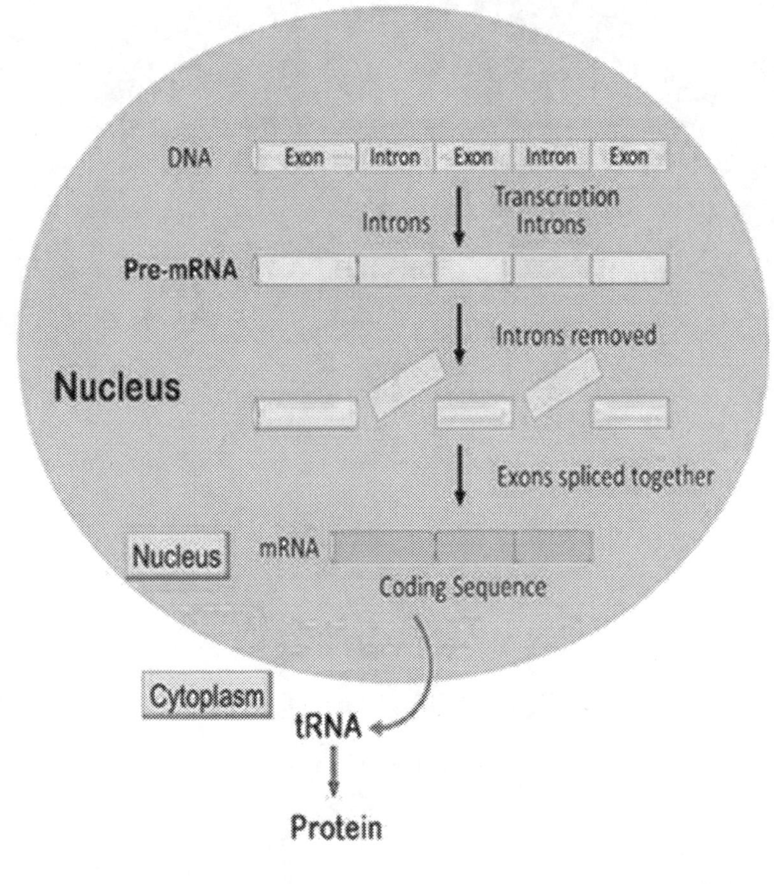

## 1.5 Gene Silencing

Silencing of genes means disruption or suppression of the expression of a gene at transcriptional or translational levels. Scientists have been working on strategies to selectively switch off specific genes in diseased tissues for the past thirty years: Some techniques have been developed to control gene expression by turning them on or off at the DNA level, "because every disease starts at the level of faulty gene expression, or viral or bacterial gene expression," said Dr. David Corey, professor of pharmacology and biochemistry at UT Southwestern Medical Center. In doing so, the research team may have paved the way for the development of new drugs designed to treat many serious diseases.

Gene silencing can be done by one or more of the following strategies:

a. Ribonuclease (RNA H) independent ODNs: Ribonuclease, which is independent of oligodeoxiribonucleotides (ODN). Ribonuclease H *(RNase H)* is a family of non-sequence-specific endonucleasees that catalyze the cleavage of RNA via a hydolic mechanism. Members of the RNase H family can be found in nearly all organisms, from bacteria toarchea to eukaryotes. Ooligodeoxynucleotides (ODNs) contains 2-aminoadenine and 2-thiothymine interact weakly with each other but form stable hybrids with unmodified complements. Oligodeoxiribonucleotides are sometimes called oligonucleotides, which is a short sequence of nucleotides (RNA or DNA) typically with twenty or fewer base pairs.
b. DNA enzymes and Ribozymes
c. RNA H dependent ODNS
d. siRNA, which is a small interfering RNA (siRNA), sometimes known as short interfering RNA or silencing RNA, is a class of double-stranded RNA molecules, 20-25 bp (base pair) in length, that play a range of roles in biology. Most notably, siRNA is involved in the RNA interference (RNAi) pathway, where it interferes with the expression of a specific gene.
e. Drugs
f. Methylation of DNA

Proteins are made out of amino acids (All contain nitrogen element) through the process of transcription of DNA to RNA and the translation fro RNA to protein. The DNA sequence in genes is copied into a messenger RNA (mRNA). Ribosomes then translate the information in the mRNA and use it to make proteins from amino acid which are carried by tRNA (transfer RNA) which enters the ribosome and meets the mRNA, Figure (13).

Figure (13): Gene silencing

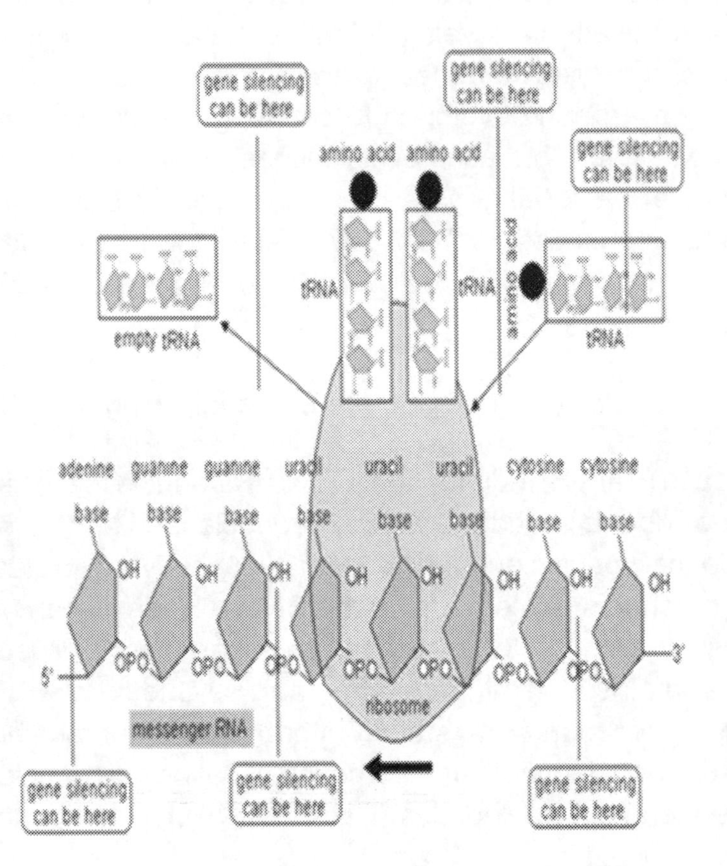


Ribosome is similar to credit card, when you swipe it, it reads the memory. It translates both tRNA and mRNA to produce protein. If interruption (silencing) to and pathway of the tRNA or mRNA, the protein will be damaged or another type of protein is produced. Protein output can be modulated to fight the disease such as AIDs or cancers. Gene silencing can also happen at the prime (5' or 3') of the DNA, since the equation of protein's output is :

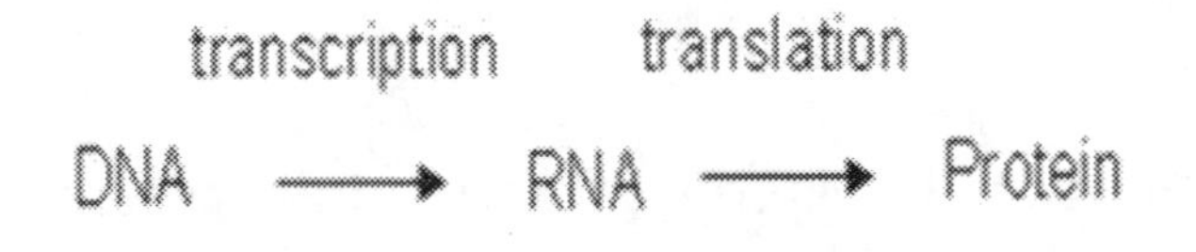


### 1.5.1 Gene silencing and aging

Science and technology have advanced at extraordinary rates in the ten years since the completion of the human genome project. Gene silencing in aging is a theory which says that aging relates to silencing of genes involved in the control of cell cycle, apoptosis, detoxification, and cholesterol metabolism.

The fact that there are considerable gene expression changes due to silencing related to aging has been confirmed in studies by different groups, conducted in yeast, worms, flies and mice. An aging mechanism is proposed based on this gene-silencing phenomenon whereby accumulation over time of methylation, oxidation, and acetylation of histones contributes to cellular senescence. Over time, this process of methylation, acetylation and oxidation of histones may lead to widespread gene silencing in diverse dividing and nondividing cell types playing part to aging of the organism.

## 1.6 Replication and Cell Division

A cell must perform DNA replication to pass down genetic material from parent to offspring. This allows for two daughter cells to e identical to the parent cell. DNA replication is the copying of DNA that occurs before cell division can take place. DNA replication faithfully duplicates the entire genome of the cell. The difference between DNA Replication and cell division is that DNA Replication is passing down genetic material, while Cell Division is when a parent cell divides to form two daughter cells.

The replication of DNA duplicates the entire genome of the cell. During DNA replication, a number of different enzymes work together to pull apart the two strands so each strand can be used as a template to synthesize new complementary strands. Each will have one original strand and one new strand. The two new daughter DNA molecules each contain one pre-existing strand and one newly synthesized strand. The following phase illustrate DNA replication: **Initiation Phase** in which two complementary strands are separated by special enzymes, **elongation phase** in which Each strand becomes a template along which a new complementary strand is built. DNA polymerase brings in the correct bases to complement the template strand, synthesizing a new strand base by base. A DNA polymerase is an enzyme that adds free nucleotides to the end of a chain of DNA, making a new double strand, and **termination phase** in which the two original strands are bound to their own, the process is finished, complementary strands, DNA replication is stopped and the two new identical DNA molecules are complete, Figure (14).

Figure (14): Replication and cell division

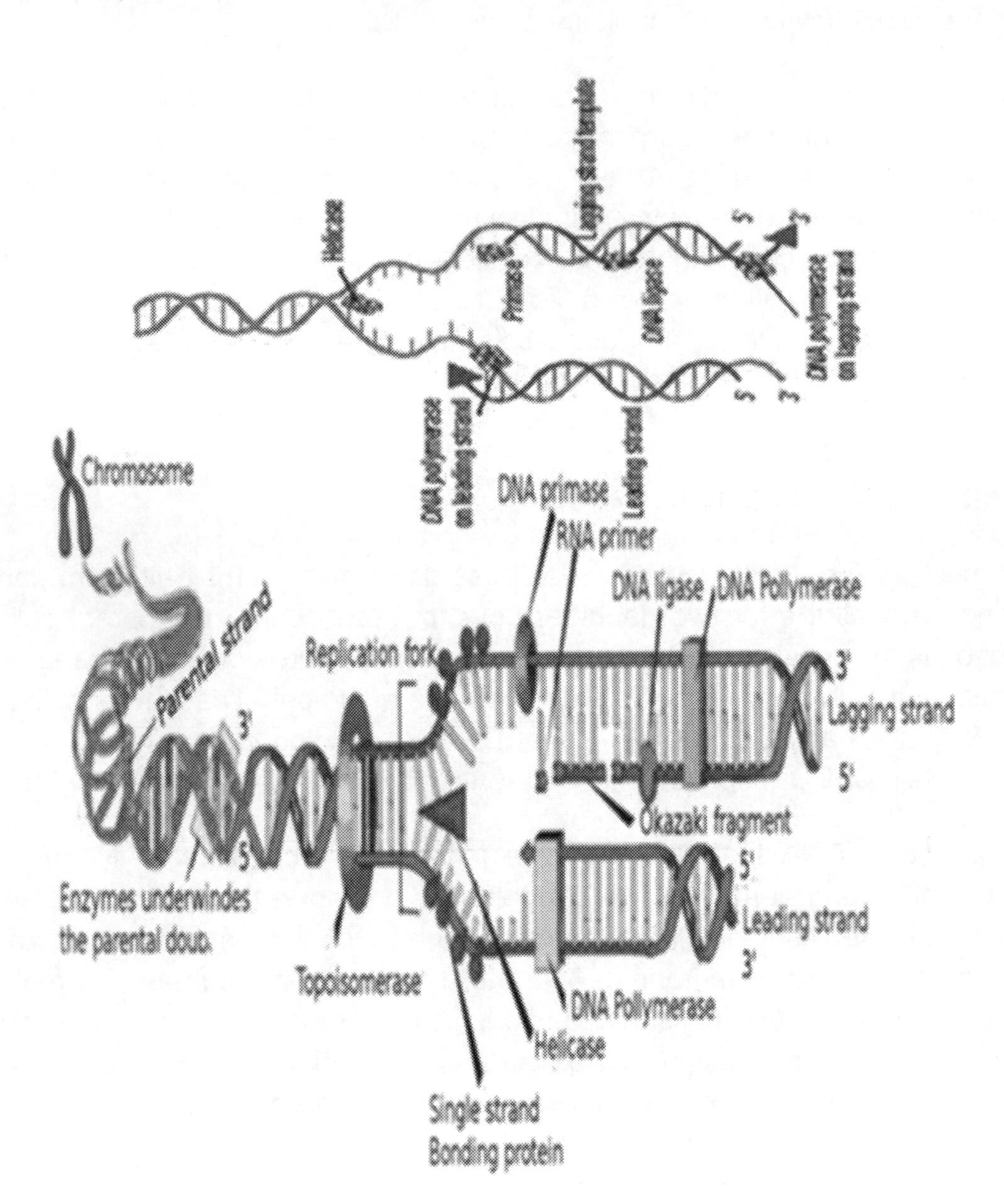


## 1.7 Sister and nonsister chromatids

Sister chromatids are alike and identical to each other. During S phase of the cell cycle the DNA is replicated and an identical copy of the chromatid is made. These two chromatids are then called sister chromatids.

Nonsister chromatids are not identical to each other as they represent unlike but homologous chromosomes. The nonsister chromatid will carry the same type of genetic information, but not exactly the same information. Some studies explained that in some species, the sister chromatids are ones responsible for DNA repair. Generally, sister chromatids cannot be different because when DNA copies, it makes an exact copy of itself; making it so that the alleles are the same.

## 1.8 Molecule Inheritance

Types of RNA: There are three types of RNA. They are mRNA, tRNA, and rRNA.

mRNA: This is called messenger RNA which includes and carries genetic information for protein synthesis from the DNA to the cytoplasm. The mRNA forms about 3 to 5% of the total cellular mRNA. The mRNA carries the message in the form of triplet codes. The life of mRNA in bacteria is about 3 minutes and in eukaryotes it lives for few hours to a few days. The mRNA is a single stranded polynucleotide chain. Each nucleotide is made up of many nucleotides. Each nucleotide has a phosphoric acid, a ribose sugar and a nitrogenous base. The nitrogenous base may be adenine, guanine, cytosine or uracil. Among RNAs, mRNA is the longest one. Most of the mRNAs contain 1000 to 15000 nucleotides. One end of the mRNA is called 5' end and other end is called 3'. At the 5' end a cap is found in most eukaryotes and animal viruses. The cap is produced by the condensation of guanylate residue. The cap helps the mRNA to bind with ribosome. The cap is followed by non-coding region. It does not contain code (message) for protein and hence it cannot translate protein. The non-coding region is followed by the initiation codon. It is made up of AUG. The initiation codon is followed by the codon region which has code for protein. It has more than 1500 nucleotides. The codon region is followed by a termination codon. It completes the translation. It is made up of UGA, UAA, or UAG in eukaryotes. The termination code is followed by non-coding region at the 3' end of mRNA. There is a polyadenylate sequence (poly A) of 200 t0 250 adenylate nucleotide (AAAAA....), Figure (15).

Figure (15): Protein coding

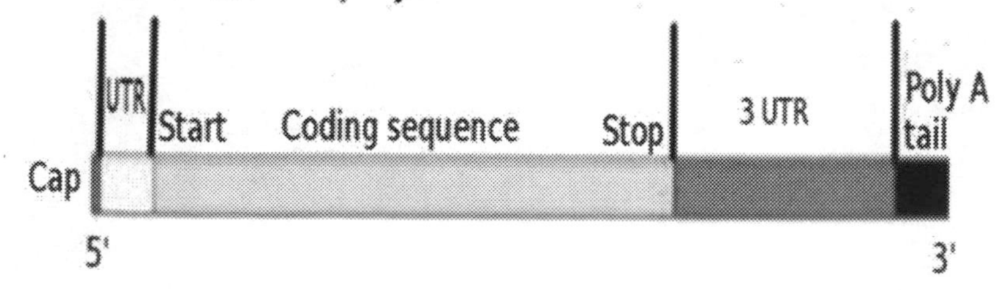

Summary:

Transcription is the part of DNA template (not the coding DNA) which contributes in transcription and consists of the followings:

1. Promoter which is the specific base sequence near the 3' end of the DNA which is recognized by RNA polymerase and then RNA pol comes and connects itself with the site and start transcription.
2. Template strand is one of the two DNA strands which are copied to form RNA.
3. Terminator is the sequence which is recognized by RNApolymerase and then transcription is stopped.

## 1.9 RNA Splicing

RNA splicing allows for the production of multiple protein isoforms from a single gene by removing introns and combining different exons Alternative splicing therefore is a process by which exons or portions of exons or noncoding regions within a pre-mRNA transcript are differentially joined or skipped, resulting in multiple protein isoforms being encoded by a single gene. Thus, alternative splicing coupled to mRNA decay can control gene expression. Splicing can be regulated so that different mRNAs can contain or lack exons, in a process called alternative splicing. Alternative splicing allows more than one protein to be produced from a gene and is an important regulatory step in determining which functional proteins are produced from gene expression. Thus, splicing is the first stage of post-transcriptional control.

Source: Boundless, Figure (16).

Figure (16): RNA splicing

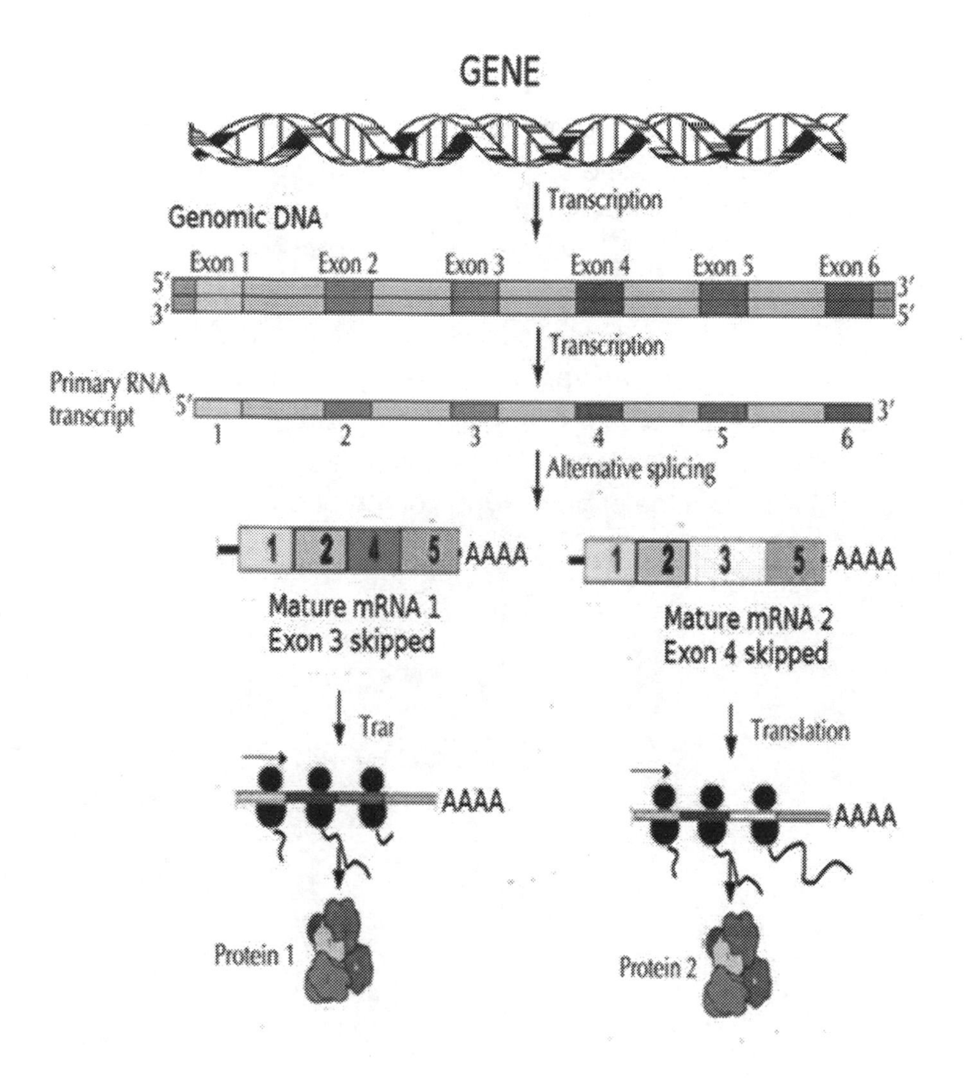


## 1.10 Gene Mutation

Some families have mutations in a gene called amyloid precurser protein (APP), which causes an abnormal form of the amyloid protein to be produced> this causes Alzheimer. These random sequences can be defined as sudden and spontaneous changes in the cell. Mutations are caused by radiation, viruses, transposons and mutagenic chemicals, as well as errors that occur during meiosis or DNA replication. They can also be induced by the organism itself, by cellular processes such as hypermutation.

One study on genetic variations between different species of Drosophila recommends that if a mutation changes a protein produced by a gene, the result is likely to be harmful, with an estimated 70 to 80 percent of amino acid polymorphisms having damaging effects, and the remainder being either neutral or weakly beneficial. Due to the damaging effects that mutations can have on genes, organisms have mechanisms such as DNA repair to avoid mutations.

Mutations can involve large sections of DNA becoming duplicated, mostly through genetic recombination. These duplications are a major source of raw material for evolving new genes, with tens to hundreds of genes duplicated in animal genomes every million years. Most genes belong to larger families of genes of shared ancestry. Novel genes are generated by several methods, commonly through the duplication and mutation of an ancestral gene, or by recombining parts of different genes to develop new combinations with new functions. Converting the information in DNA into protein is a two-step process, involving transcription and translation. In transcription each mRNA nucleotide pairs with the complementary DNA nucleotide. In translation, each tRNA nucleotide pairs with the complementary mRNA nucleotide. Thus, a variation in the DNA sequence can vary and change the amino acid sequence of the protein. There are three basic types of mutations: insertion, deletion and substitution. Some mutations are silent, meaning that there is no change in the protein, while others can initiate major changes in the protein. Types of mutations are shown in Figure (17).

Figure (17) Gene mutation

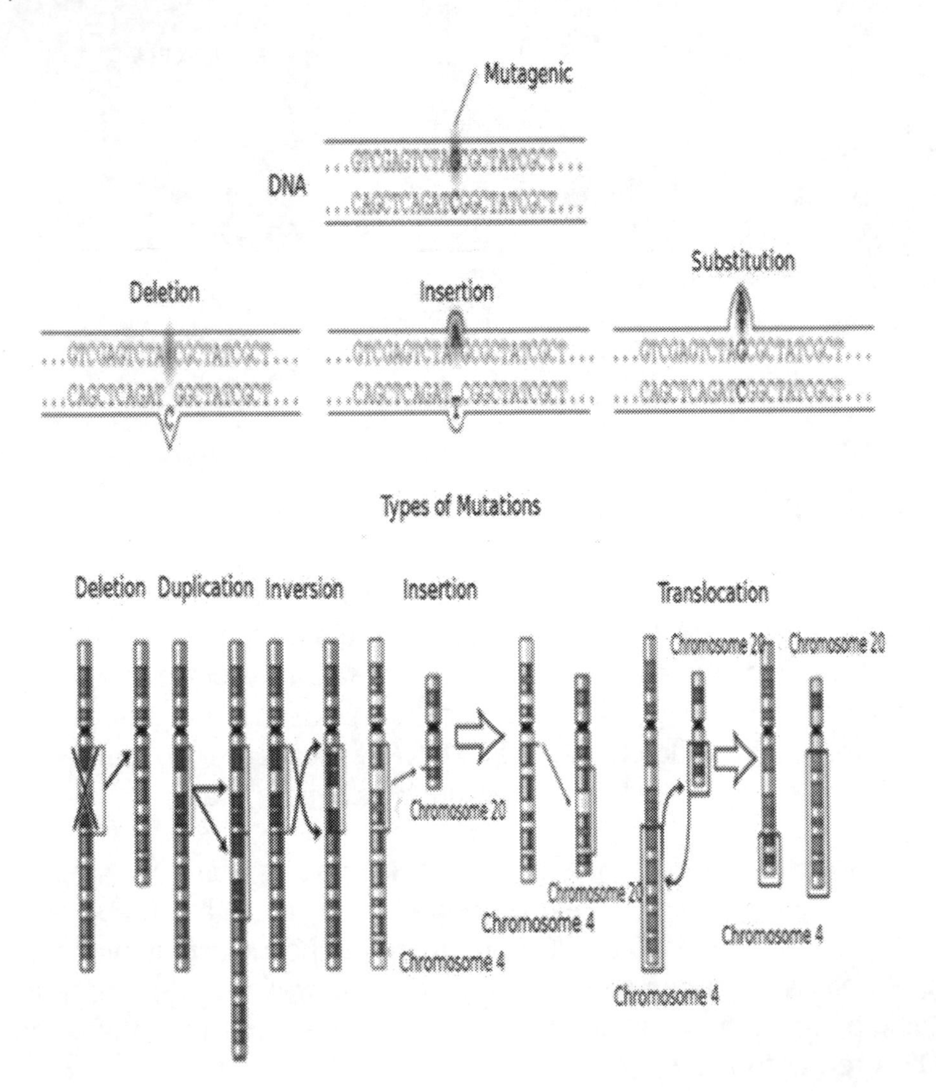

Gene mutations can be classified in two major ways:

- Heredity mutations are inherited from a parent and are present throughout a person's life in almost every cell in the body. These mutations are also called germ line mutations because they are present in the parent's egg or sperm cells, which are also called germ cells. When a sperm and an egg cell unite, the resulting fertilized egg cell receives DNA from both parents. If this DNA has a mutation, the offspring that grows from the fertilized egg will have the mutation in all human cells.

Somatic mutations are frequently caused by environmental factors, such as exposure to ultraviolet radiation or to certain chemicals. Somatic mutations occur at some time during a person's life and are present only in certain cells, not in every cell in the body. Acquired mutations in somatic cells cannot be passed on to the next generation.

### 1.10.1 Causes

Two reasons of mutations are spontaneous mutations and induced mutations.

Spontaneous mutation

This can be caused by:

- Denaturation of the new strand from the template during replication, followed by renaturation in a different spot ("slipping"). This can lead to insertions or deletions
- Tautomerism – A base is changed by the repositioning of a hydrogen atom, altering the hydrogen bonding pattern of that base resulting in incorrect base pairing during replication.
- Depurination – Loss of a purine base (A or G) to form an apurinic site
- Deamination – Hydrolysis changes a normal base to an atypical base containing a keto group in place of the original amine group. Examples include C → U ·

Induced mutation

Induced mutations on the molecular level can be caused by:

Chemicals

- Oxidative damage

·Nitrous acid converts amine groups on A and C to diazo groups, altering their hydrogen bonding patterns which leads to incorrect base pairing during replication.

Diazo refers to a type of organic compound called diazo compound that has two linked nitrogen atoms (azo) as a terminal functional group; using functionalizations of O–H and N–H bonds,[3] cross-coupling with diazocompounds.

Radiation

- Ultraviolet radiation (nonionizing radiation). Two nucleotide bases in DNA – cytosine and thymine – are most vulnerable to radiation that can change their properties. Ultraviolet light induces the formation of covalent linkages by reactions localized on the C=C double bonds. UV light can induce adjacent pyrimidine bases in a DNA strand to become covalently joined as a pyrimidine dimer. UV radiation, particularly longer-wave UVA, can also cause oxidative damage to DNA. Mutation rates also vary across species.

## 1.11 Gene Expression

1. Gene expression is the process by which information from a gene is used in the synthesis of a functional gene product. These products are often proteins, but in non-protein coding genes such as transfer RNA (tRNA) genes, the product is a functional RNA, Figure (14).
2. DNA is the blueprint of every cell • Proteins control cell shape, reproduction, function, and synthesis of biomolecules • The information in DNA genes must therefore be linked to the proteins that run the cell • Proteins are the "molecular workers" of the cell.
3. The base sequence in a DNA gene dictates the sequence and type of proteins (amino acids) in translation • Bases in mRNA are read by the ribosome in triplets called codons • Each codon specifies a unique amino acid in the genetic code. Each mRNA initiates a start and a stop codon.
4. Transcription of a DNA gene into RNA has three phases – Initiation – Elongation – Termination.
5. Initiation phase of transcription 1. DNA molecules are unwound and strands are separated at the beginning of the gene sequence 2. RNA polymerase binds to promoter region at beginning of a gene on template strand.
6. Polymerase of RNA synthesizes a sequence of RNA nucleotides along DNA template strand. Bases in newly synthesized RNA strand are complementary to the DNA template strand. RNA strand peels away from DNA template strand as DNA strands repair and wind up.
7. Elongation • As elongation proceeds, one end of the RNA drifts away from the DNA; RNA polymerase maintains the other end temporarily attached to the DNA template strand.
8. Termination – RNA polymerase reaches a termination sequence and releases completed RNA strand.

9. Translation Translation is the process by which ribosomes interprete the genetic message in the mRNA and produce a protein product according to the message's instruction..

Aminoacyl tRNA synthetase enzyme which activates Elongation factors, Initiation factors, termination factors, Amino acids, mRNA, tRNA, Ribosomes, and requirements of Translation, Figure (18).

Figure (18): Gene expression

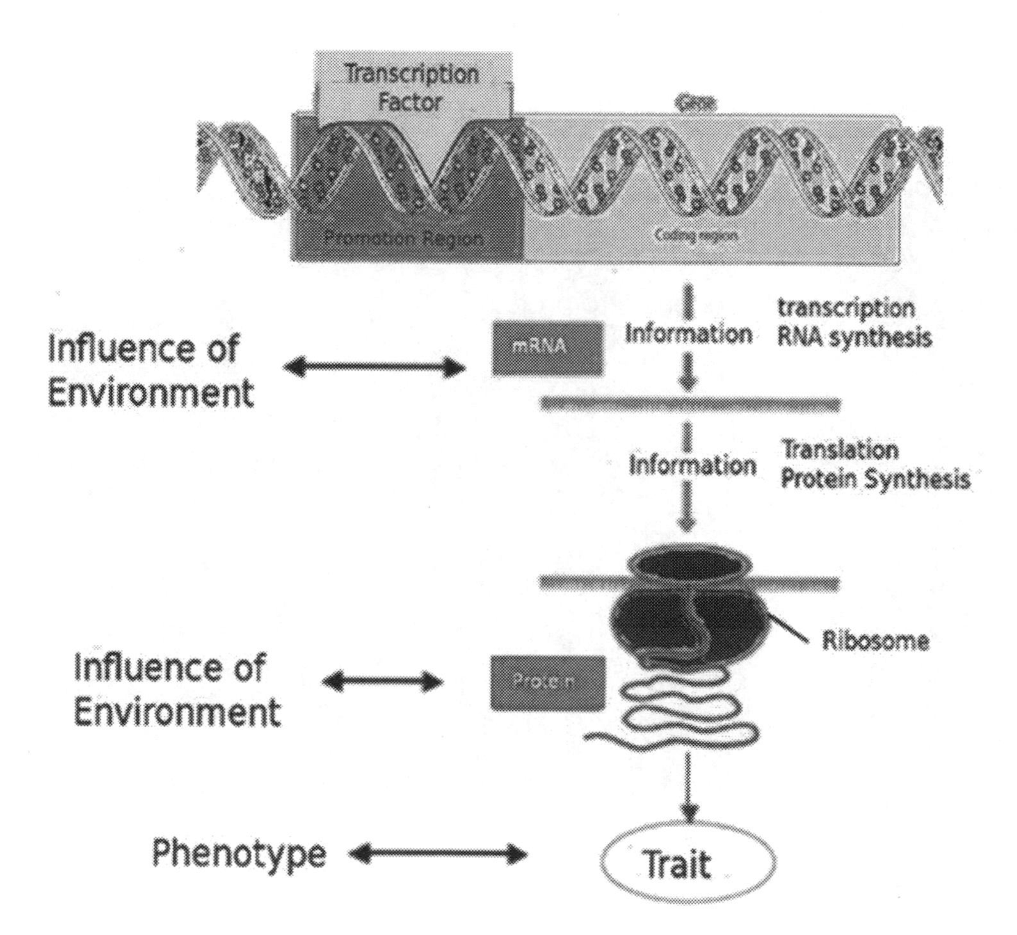


## 1.12 Genetic Coding

With four possible bases, the three nucleotides can give $4^3 = 64$ different possibilities of amino acids, and these combinations are used to specify the 20 different amino acids used. The genetic code was finally "discovered" in 1966. Marshall Nirenberg, Heinrich Mathaei and Severo Ochoa demonstrated that a sequence of three nucleotide bases, a codon or triplet, determines each of the amino acids (20) found in nature.

The ribonucleic acid (RNA) that is directly involved in the transcription of the pattern of bases from the DNA to provide a blueprint for the construction of proteins is called messenger RNA or mRNA. The pattern for protein synthesis is then read and translated into the language of amino acids for protein construction with the help of ribosome and transfer RNA.

Methionine is the only amino acid specified by just one codon, AUG. The stop codons are UAA, UAG, and UGA. They encode no amino acid. The ribosome pauses and falls off the mRNA. The stretch of codons between AUG and a stop codon is called an open reading frame (ORF). Computer analysis of DNA sequence can calculate the existence of genes based on ORFs. Other amino acids are specified by more than one codon--usually differing at only the third position. Translation engages the conversion of a four base code of the DNA (ATCG) into twenty different amino acids. A codon or triplet of bases specifies a given amino acid. Most amino acids are specified by more than one codon.

The conversion of codon information into proteins is conducted by transfer RNA. Each transfer RNA (tRNA) has anticodon which can base pair with a codon, Figure (19).

Figure (19): Conversion of codons into proteins

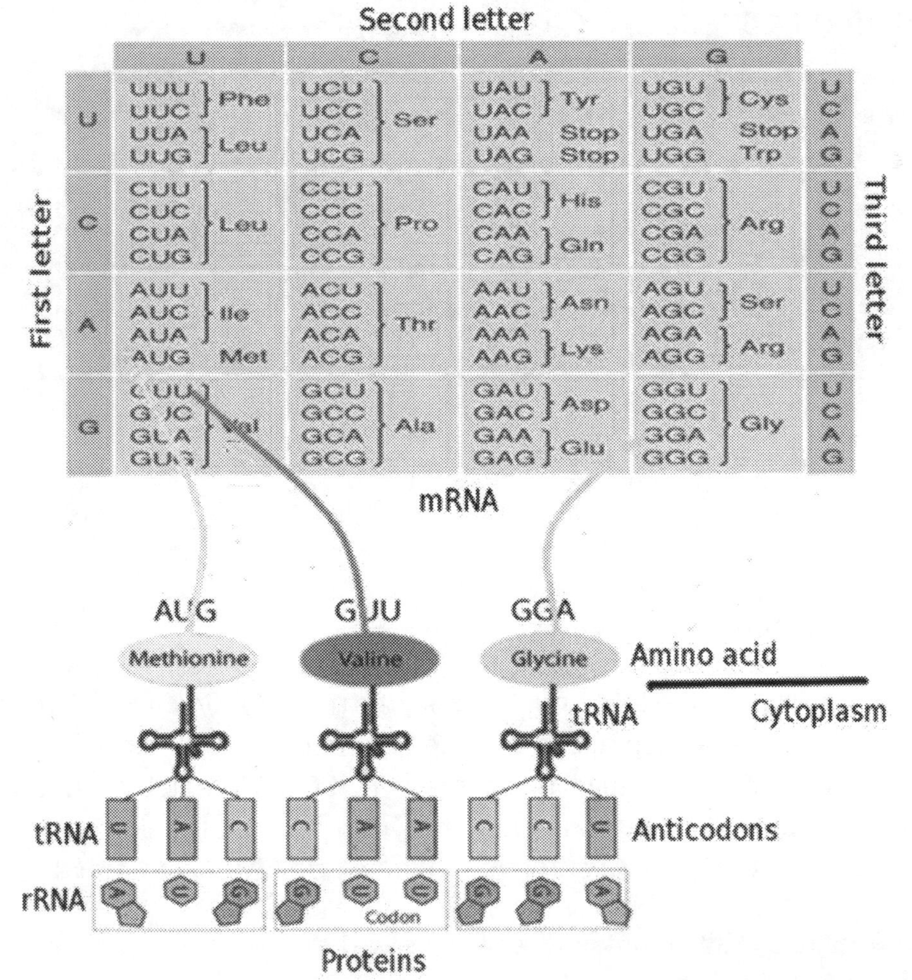

## 1.13 Nucleic Acid Hybridation

Most techniques of eukaryotic gene chemical analysis are based on nucleic acid hybridization.

DNA Hybridization base pairing can be of two single strands of DNA or RNA or DNA-DNA.

This technique involves annealing single-stranded pieces of RNA and DNA to allow complementary strands to form double-stranded hybrids. If DNA is cut into small pieces and each piece dissociated into two single strands in the solution, each strand should find and reunite with its complementary partner, given sufficient time. Similarly, RNA synthesized from a particular region of DNA would be expected to bind to the strand from which it was transcribed.

Though a double-stranded DNA sequence is generally stable under physiological conditions, changing these conditions in the laboratory (generally by raising the surrounding temperature) will cause the molecules to separate into single strands. The hybridization can be carried out in solution or with one component immobilized on a gel or, most commonly, on nitrocellulose paper. Cutting DNA can be done by radioactive DNA precursors, and the viral enzyme reverse transcriptase (endonucleases). Fluorescence in situ hybridization (FISH) is a laboratory method used to detect and locate a DNA sequence, often on a particular chromosome.

Hybridizations are done in all combinations: DNA-DNA (DNA can be turned into single-stranded by heat penetration), DNA-RNA or RNA-RNA.

In situ hybridization involves hybridizing a marked nucleic acid (often labeled with a fluorescent dye) to suitably prepared cells or histological sections. DNA hybridization is shown in Figure (20).

Figure (20): DNA hybridization

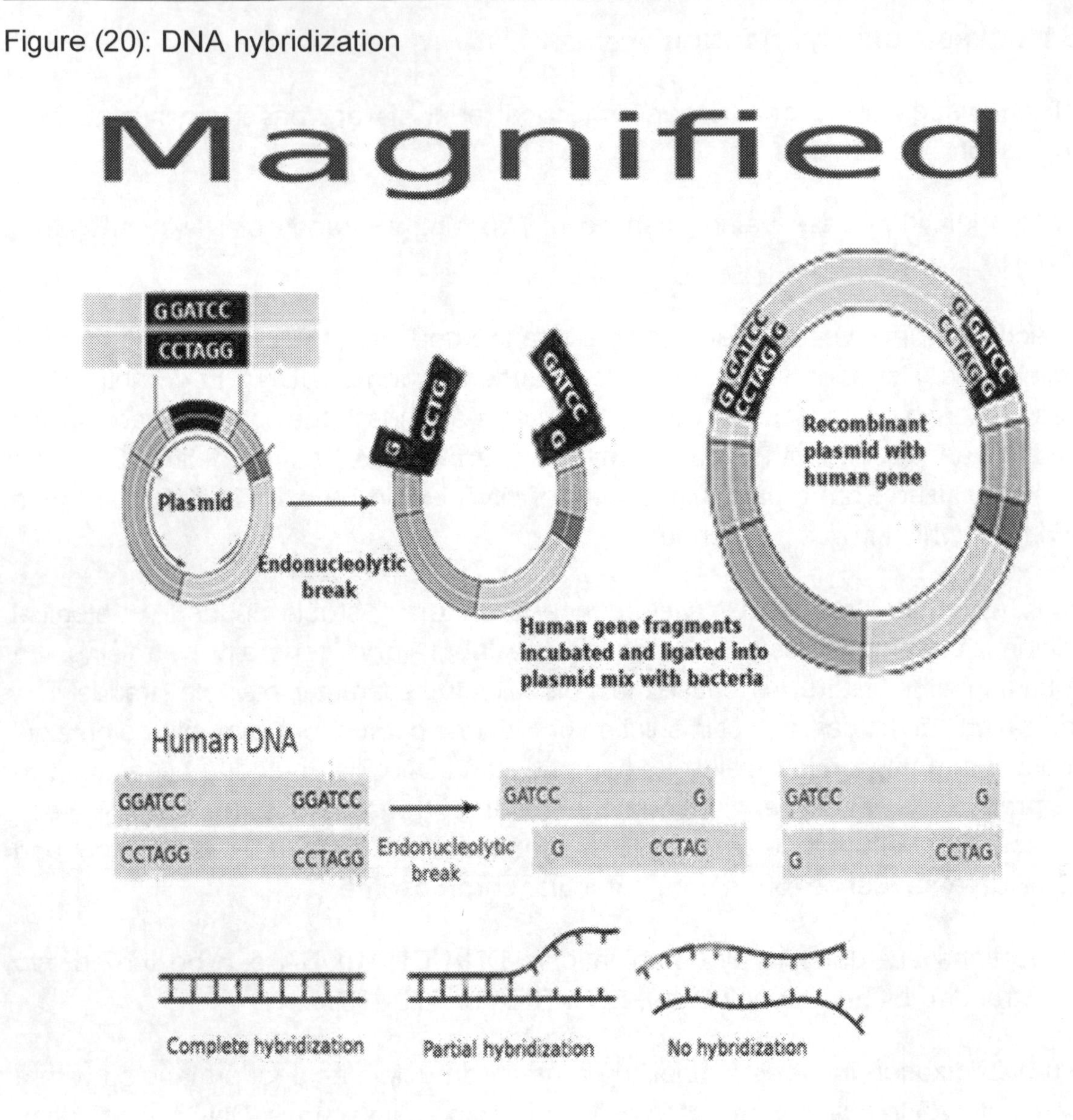


### 1.13.1 Monohybrid and dihybrid cross

A monohybrid cross considers only a single character whereas a dihybrid cross considers two characters (and a trihybrid cross would consider three characters, and so on). For example, When crossing 2 heterozygotes in a monohybrid cross involving a character that follows complete dominance, the Mendelian genotypic and phenotypic ratios are:

1:2:1 (1 homozygous dominant: 2 heterozygous: 1 homozygous recessive)

For example, if you cross pure-breeding tall pea plants with pure-breeding short pea plants, you are considering only 1 character: stem length. That is a monohybrid cross. If you cross pure-breeding tall, round-seed pea plants with pure-breeding short, wrinkled-seed pea plants, then you are considering 2 characters in the cross and it is a dihybrid cross, Figure (21).

Figure (21): Monohybrid and dihybrid cross

Monohybrid cross

Father is heterozygous for particular trait (Aa).

Mother is also heterozygous for the same trait (Aa).
Homozygous dominant (AA) =1/4
Heterozygous (Aa) = 1/2
Homozygous recessive = 1/4

| ♀ \ ♂ | A | a |
|---|---|---|
| A | AA | Aa |
| a | Aa | aa |

Dihybrid cross
A and a represent one trait, and B and b represent a different trait that is linked to inheritance of A or a.

Dominant for A and B = 9/16
Dominant for A, recessive for b = 3/16
Recessive for a, dominant for B = 3/16
Recessive for a, recessive for b = 1/16

| | AB | Ab | aB | ab |
|---|---|---|---|---|
| AB | AABB | AABb | AaBB | AaBb |
| Ab | AABb | AAbb | AaBb | Aabb |
| aB | AaBB | AaBb | aaBB | aaBb |
| ab | AaBb | Aabb | aaBb | aabb |

## 1.14 Gene/Chromosome Abnormalities

A chromosome anomaly, abnormality, aberration, or mutation is a missing, extra, or irregular portion of chromosomal DNA. It can be from an atypical number of chromosomes or a structural abnormality in one or more chromosomes.

- Germline abnormalities occur in gametes. These abnormalities are especially significant because they can be transmitted to offspring and every cell in the offspring will have the mutation.
- Somatic abnormalities occur in other cells of the body. These abnormalities may have little effect on the organism because they are confined to just one cell and its daughter cells. Somatic mutations cannot be passed on to offspring.

Abnormalities also differ in the way that the genetic material is changed. Mutations may change the structure of a chromosome or just change a single nucleotide. Possible ways these abnormalities can arise are illustrated in Figure (22).

Figure (22): Gene abnormalities

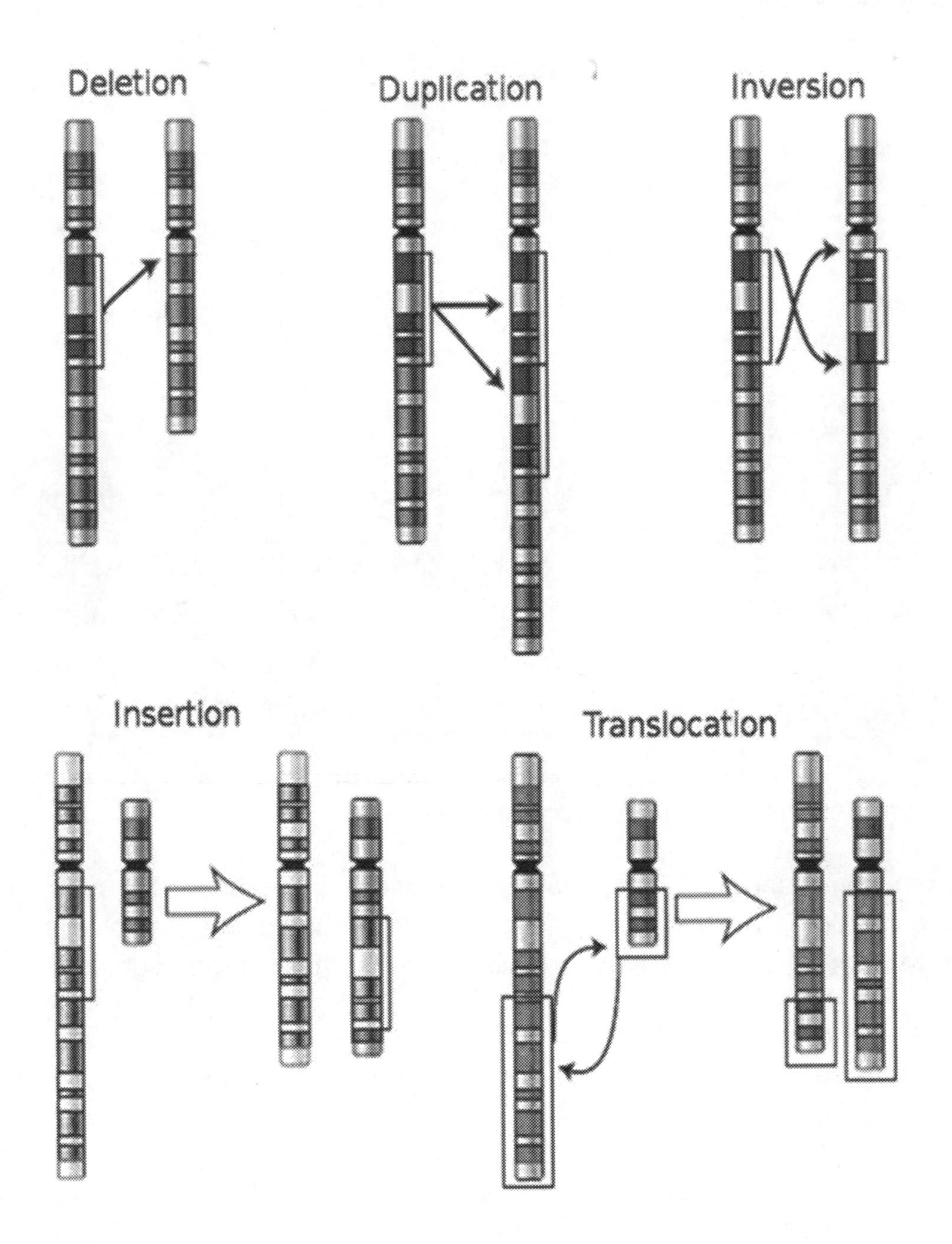


### 1.14.1 Genetic Disorders

To understand how genetic diseases come about, remember that all genes (except those in egg and sperm cells) occur in pairs. One member of each pair comes from the father and one from the mother.

Most common disorders are:

- P – Point mutation, or any insertion/deletion entirely inside one gene
- D – Deletion of a gene or genes
- C – Whole chromosome extra, missing, or both
- T – Trinucleotide repeat disorders: gene is extended in length

Inherited diseases are of two types. The first are those resulting from a disease-causing dominant gene inherited from one of the parents. In this case, one of the parents who passes on the unhealthy gene must also be a suffer of the disease. The second type of genetic disease comes about when a child receives the same disease-causing recessive gene from both parents. Since there is now no choice but for one of the unhealthy genes to be turned on, the child will suffer from the disease. Cystic fibroses is caused in this way, Figure (23).

Figure (23): Genetic disorder

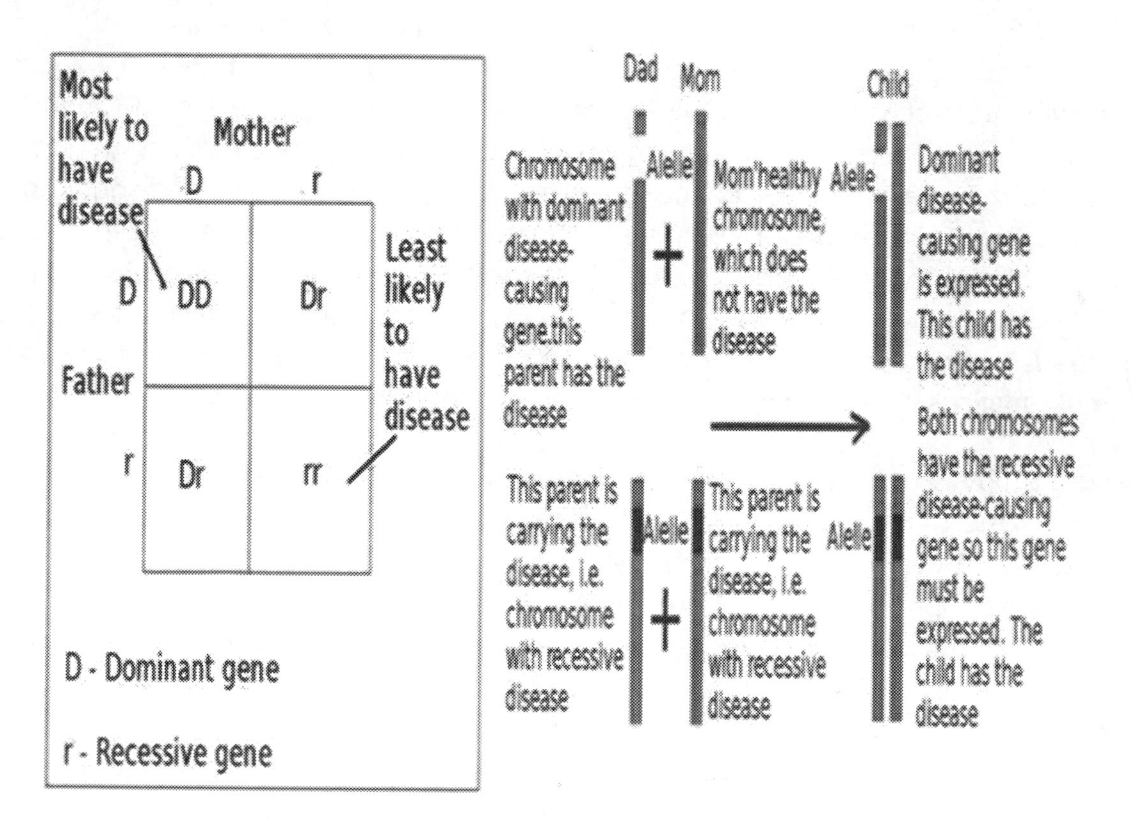


## 1.15 Gene Therapy

Gene therapy is an experimental technique that uses genes to treat or inhibit disease. In the future, this technique may allow doctors to reverse aging by introducing a gene into aged individuals to rejuvenate cells. Researchers are testing several approaches to gene therapy:

- Replacing old genome with young ones.
- Inactivating, or isolate a mutated gene that is malfunctioning.
- Introducing a new gene into the body to help fight a disease.
- Replacing a mutated gene that causes disease with a healthy copy of the gene.

Gene therapy is a promising field that offers fundamentally new ways of curing human illness. Genes, which are carried on chromosomes, are the basic physical and functional units of heredity. There are two types of genes: Germ line gene, i.e., sperm or eggs, are modified by the introduction of functional genes, which are ordinarily integrated into their genomes. Therefore, the change due to therapy would be heritable and would be passed on to offsprings, and the Somatic gene therapy in which the therapeutic genes are passed on into the somatic cells of a patient. Any modifications and effects will be restricted to the individual patient only, and will not be inherited by the patient's offspring.

Gene therapy is the introduction of genetic material into cells for therapeutic purposes. Recent scientific breakthroughs in the genomics field and our understanding of the important role of genes in disease has made gene therapy one of the most rapidly advancing fields of biotechnology with great promise for treating inherited and acquired diseases. Many jurisdictions and ethical principles prohibit this for exercising in human beings, for a variety of technical and ethical reasons. Genes are specific sequences of bases (nucleotides) that program instructions on how to synthesize enzymes and proteins. Although genes get a lot of attention, it's the proteins that perform most life functions and even make up the majority of cellular structures. When genes are changed so that the encoded enzymes and proteins are unable to perform their normal functions, diseases and aging are the consequences.

Gene therapy is being thoroughly examined as a cure for several genetic diseases. Out of all the genetic disorders, gene therapy for both sickle cell and hemophilia diseases has the most beneficial characteristics for this potential cure. Gene therapy works in hemophilia by using DNA as the drug and viruses as the deliverer. For example, a virus containing the gene that produces Factor VIII or Factor IV (in case of Hemophilia B) is injected into a large group of cells in the patient. The hope of gene therapy is to have the cell produce more of the cured cells and proliferate throughout the rest of the body. If successful, the patient would never need factor replacement therapy again and would be cured of diseases. Figure (24) indicates the process of gene therapy injection.

Figure (24): Process of gene therapy

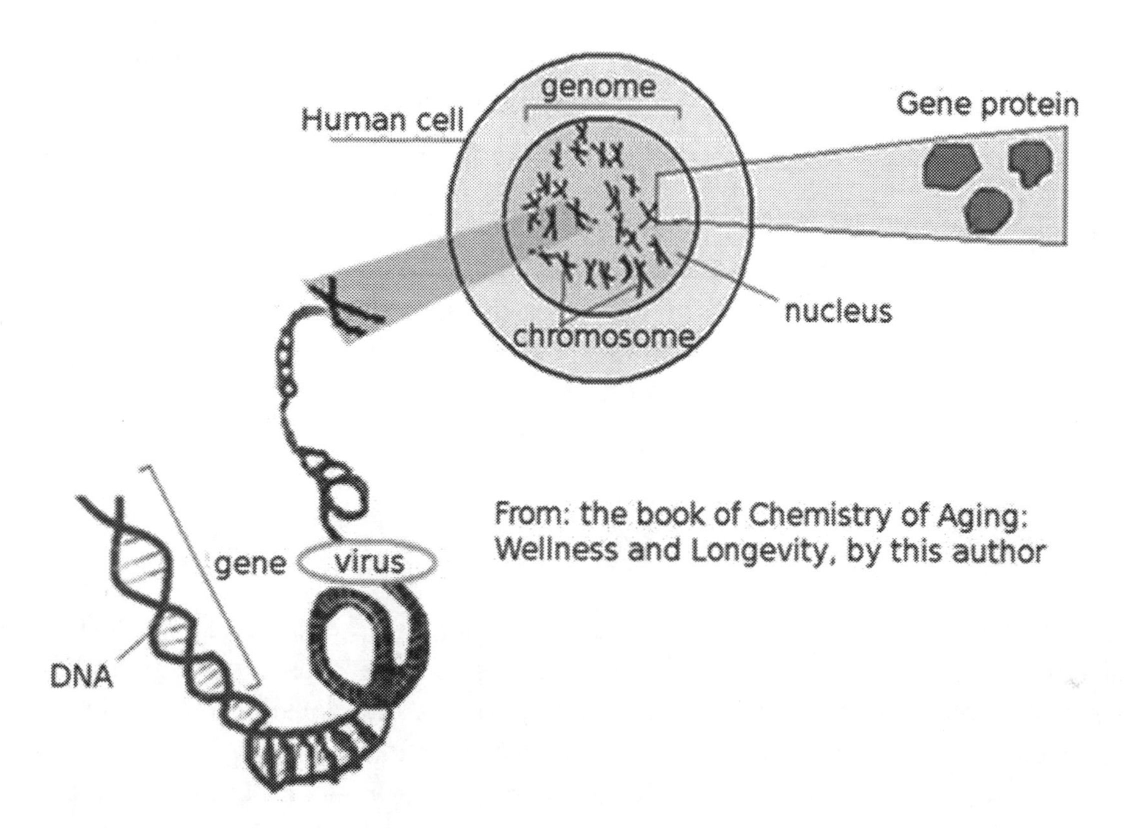


Pioneering gene therapy has brought back some vision to patients with a rare form of genetic blindness for as long as four years, raising hopes it could be used to cure common causes of vision loss, new University of Oxford research claims.

Pennsylvania researchers using gene therapy have made major improvements in restoring vision in 12 patients with a rare inherited visual defect.

### 1.15.1 Gene therapy and aging

Some studies show that it is possible to develop a telomerase-based anti-aging gene therapy without increasing the incidence of cancer. Aged organisms accumulate damage in their DNA due to telomere shortening, [this study] finds that a gene therapy based on telomerase production can repair or delay this kind of damage," they add.

It is believed that telomere length was a primary aging clock in humans, but not in mice.

Some researchers in this area, and "life extensionists", "immortalists" or "longevists" (those who wish to achieve longer lives themselves), believe that future breakthroughs in tissue rejuvenation, stem cells, regenerative medicine, molecular repair, gene therapy, pharmaceuticals, and organ replacement will eventually enable humans to have indefinite lifespans through complete rejuvenation to a healthy youthful condition. Studies have also shown that the effects of adding one or multiple copies of various genes, that leads to the increased expression of their gene products, has resulted in the extension of life span in model systems such as worms, fruitflies, rodents and cultured cells. Some such transgenic manipulations include the addition of gene(s) such as antioxidant genes superoxide dismutase (SOD) and catalase, NAD+-dependent histone deacetylases sirtuins, forkhead transcription factor FOXO, heat–shock proteins (HSP), heat–shock factor, protein repair methyltransferases and klotho, which is an inhibitor of insulin and IGF-1 signaling. Another system in which genetic interventions have been tested is the Hayflick system of limited proliferative life span of normal diploid differentiated cells in culture.

### 1.15.2 Hayflick limit and gene therapy

The Hayflick limit is the number of times a normal human cell population will divide until cell division stops. Empirical evidence shows that the telomeres associated with each cell's DNA will get slightly shorter with each new cell division until they shorten to a critical length. Hayflick demonstrated that a population of normal human fetal cells in a cell culture will divide between 40 to 60 times. The Hayflick Limit, which places the maximum potential lifespan of humans at 120, the time at which too many cells can no longer split and divide to keep things going.

To conclude, Hayflick stated that Some types of cells, such as those that produce red and white blood corpuscles, can divide millions of times. Others, such as most nerve cells, do not reproduce at all. If a cell's Hayflick limit is 50, for instant, it will divide 50 times and then become senescent. It withers and dies. When enough of our cells die, we die.

Leonard Hayflick and Paul Moorhead found out that human cells derived from embryonic tissues can only divide a finite number of times in culture. They divided the stages of cell culture in three phases: Phase I is the primary culture, when cells from the explants simply multiply to envelop the surface of the culture flask. Phase II represents the period when cells divide in culture. Briefly, once cells cover a flask's surface, they stop multiplying. For cell growth to carry on, the cells must be sub cultivated. To do so, one removes the culture's medium and adds a digestive enzyme called trypsin that dissolves the substances keeping cells together. If you add growth medium afterwards, you get the cells in suspension that can then be divided by two--or more--new flasks. Later, cells attach to the flask's floor and start dividing once again until new sub cultivation is required. Cells divide energetically and can often be subcultivated in a matter of a few days. Eventually, however, cells start dividing slower, which marks the beginning of Phase III. Eventually they stop dividing at all and may or not die. Hayflick and Moorhead noted that cultures stopped dividing after an average of fifty cumulative population doublings. This phenomenon is known as Hayflick's limit, Phase III phenomenon, or, as it will be called herein, replicative senescence as per Figure (24).

Figure (24): Phases of replication of cell culture

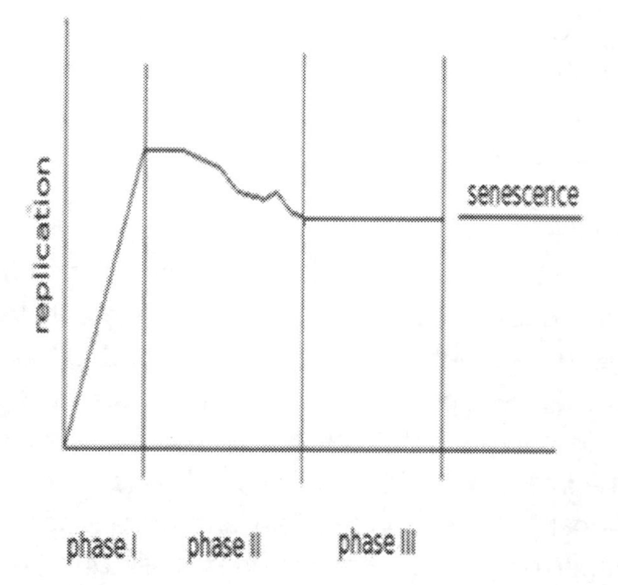


Hayflick and Moorhead used to work with fibroblasts, a type of cell found in connective tissue; produces collagen, but RS (replicative senescence) has been found in other cell types: keratinocytes (epidermal cells), endothelial cells (endothelium cells are the fine layers of cells that line the interior surface of blood vessels), lymphocytes (a lymphocyte is a type of white blood cell in the immune system), vascular smooth muscle cells, adrenocortical cells, chondrocytes (are the only cells found in cartilage), etc. In addition, RS is observed in cells derived from embryonic tissues, in cells from adults of all ages, and in cells taken from many animals: mice, Galapagos tortoises, chickens, etc. Early results suggested a relation between the number of CPDs (cumulative population doublings) cells go through in culture and the longevity of the species from which the cells were derived. For example, cells from the Galapagos tortoise, which can live over a century, divide about 110 times, while mouse cells divide roughly 15 times. In addition, cells taken from patients with progeroid syndromes such as Werner syndrome endure far less CPDs than normal cells. Exceptions exist and certain cell lines never reach RS. Some cells have no Hayflick limit. Barring trauma from outside, they are immortal. They can be died, but they do not age. The "lowly" bacteria are immortal. They can be killed – by heat, starvation, radiation, lack of water, or being eaten by another organism. But they do not age. Bacteria keep on dividing forever, until some outside agency kills them. Cancer cells are similarly immortal. They keep on dividing and dividing, eternally, unless they are killed or their host dies. "HeLa" cells, taken from the tumor of Henrietta Lacks in 1951 (Henrietta Lacks (August 18 (?), 1920 – October 4, 1951) was the involuntary donor of cells from her cancerous tumor, which were cultured by George Otto Gey to create an immortal cell line for medical research. This is now known as the Hela cell line), are still reproducing as vigorously as they did nearly 50 years ago. Human germline cells -- ova and sperm cells -- also show no Hayflick limit.

Senescent cells are growth arrested in the transition from phase G1 to phase S of the cell cycle. The growth arrest in RS is irreversible in the sense that growth factors cannot stimulate the cells to divide, even though senescent cells can remain metabolically active for long periods of time.

Bone marrow transplant is the most widely used stem-cell therapy, but some therapies derived from umbilical cord blood are also in use. Research is underway to develop various sources for stem cells, and to apply stem-cell treatments for neurodegenerative diseases and conditions, diabetes, heart disease, and other conditions. Recent studies showed that marrow cells (Marrow stromal cells, MSCs) may be used to repair senescence cells and gene therapy. If marrow cells are to be used for cell and gene therapy, it will be important to define the conditions for isolation and expansion of the cells. As demonstrated by Friedenstein and colleagues, MSCs are relatively easy to isolate from marrow from most species by their adherence to tissue culture plates and flasks. However, the cells display several unusual features as they expand in culture. The difficulty of carrying out experiments in animal models with MSCs and other marrow cells has prompted scientists to develop a coculture system to study the repair of injured cells and tissues by MSCs. In initial experiments, MSCs were cocultured with heat-shocked human small airway epithelial cells. Figure (25) is a schematic showing MSCs with gene green fluorecent protein injected into senescence cells to produce repaired cells.

Figure (25): Using MSCs as gene therapy

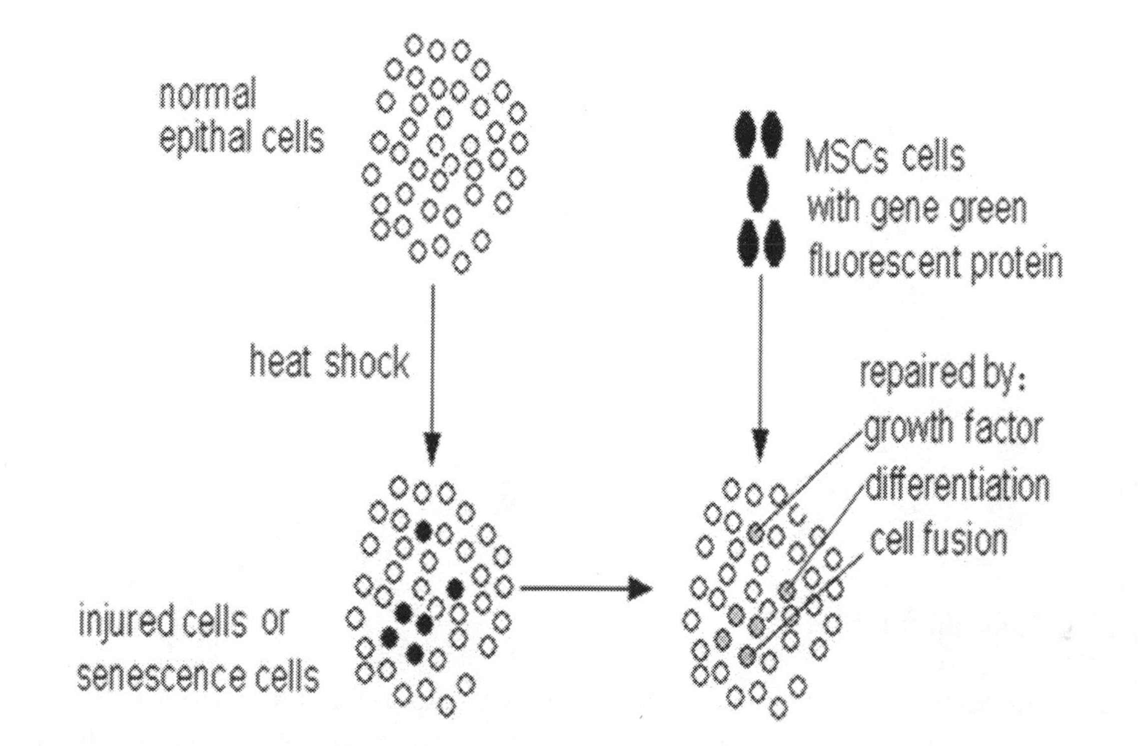


## 1.16 Traits

The combination of the two alleles is called a genotype. One allele is from your mother and one from your father, and these genes can be slightly different. In some cases, the gene is dominant, which means that the variant of the trait that it is responsible for will take over. Our genotype determines our measurable physical characteristics known as phenotypes, or traits. Take the ABO blood type as an example. Blood type is an example of a common multiple allele trait. There are 3 different alleles for blood type, (A, B, & O). A is dominant to O when alleles are A and O. B is also dominant to O when alleles are B AND O. A and B are. If you have one allele for blood type A and one allele for blood type O, then your genotype is AO and your phenotype is blood type A.

The A and B antigen molecules on the surface of red blood cells are made by two different enzymes. These two enzymes are encoded by different versions, or alleles, of the same gene. The A allele codes for an enzyme that makes the A antigen, and the B allele codes for an enzyme that makes the B antigen. A third version of this gene, the O allele, codes for a protein that is not functional; it makes no surface molecules at all. Everyone inherits two alleles of the gene, one from each parent. The combination of your two alleles determines your blood type.

Figure (26): Phenotypes based on blood type

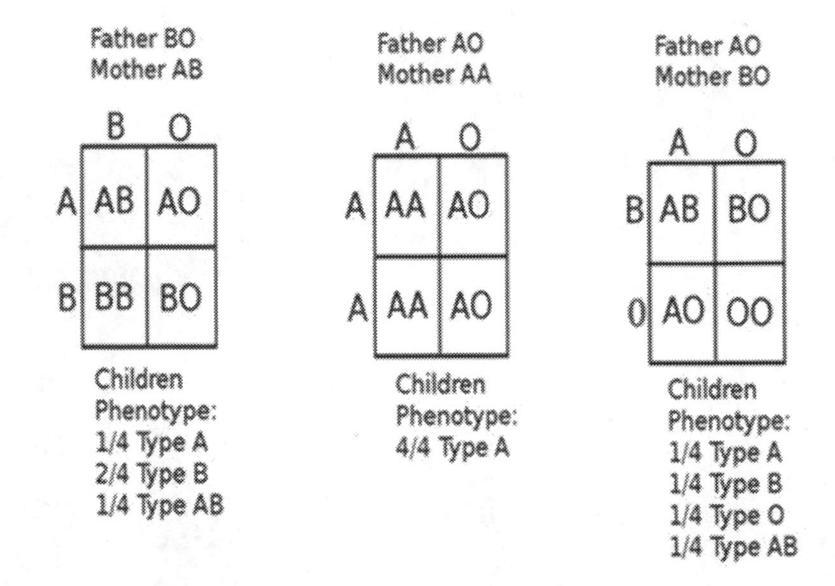


### 1.16.1 Mendelian Inheritance

Mendel considered in his experiments on peas plants traits such as color, tall, shape and other visible traits. He did not account for hidden characters and traits such as taste, emotion, intelligence, age, and other factors dominating animal behavior. Offspring may or may not be identical physical copies of parents; rather they are the combination of all characters and traits whether visible or invisible, sensible or insensible. Let's take, for example the first law of Mendel, Table (1).

Table (1): Genotypes crossing

| | Y | y |
|---|---|---|
| Y | YY | Yy |
| y | Yy | yy |

Three genotypes have traits influenced by the dominant phenotype in parents. If Y represents the tall and y represent the short, then three genotypes are tall, and one is short. Expand above to include two traits in each parent:

Consider:

Y: dominant allele for tall
y: recessive allele for short
W: Dominant allele for white
w: recessive allele for black

The output as per Mendel would be combination of YyWw x YyWw, Table (2).

Table (2): Crossing between dominant and recessive alleles

| | YW | Yw | yW | yw |
|---|---|---|---|---|
| YW | YWYW (tall) | YWYw (tall) | YWyW (tall) | YWyw (tall) |
| Yw | YwYW (tall) | YwYw (tall) | YwyW (tall) | Ywyw (tall) |
| yW | yWYW (tall) | yWYw (tall) | **yWyW (short)** | **yWyw (short)** |
| yw | ywYW (tall) | ywYw (tall) | **ywyW (short)** | **Ywyw (short)** |

Note that there are twelve tall, and only four shorts.

For simplicity, Alphabetical and Numerical figures will be used.

Let us expand the alleles to include a third character dwarf and brown. Table (3) would be a combination of ABC123, and still the dominant is A in the father and 1 in the mother, Table (3).

Table (3): Expanding traits

| | A1 | A2 | A3 | B1 | B2 | B3 | C1 | C2 | C3 |
|---|---|---|---|---|---|---|---|---|---|
| A1 | A1A1 | A1A2 | A1A3 | A1B1 | A1B2 | A1B3 | A1C1 | A1C2 | A1C3 |
| A2 | A2A1 | A2A2 | A2A3 | A2B1 | A2B2 | A2B3 | A2C1 | A2C2 | A2C3 |
| A3 | A3A1 | A3A2 | A3A3 | A3B1 | A3B2 | A3B3 | A3C1 | A3C2 | A3C3 |
| B1 | B1A1 | B1A2 | B1A3 | B1B1 | B1B2 | B1B3 | B1C1 | B1C2 | B1C3 |
| B2 | B2A1 | B2A2 | B2A3 | B2B1 | **B2B2** | **B2B3** | B2C1 | **B2C2** | **B2C3** |
| B3 | B3A1 | B3A2 | B3A3 | B3B1 | **B3B2** | **B3B3** | B3C1 | **B3C2** | **B3C3** |
| C1 | C1A1 | C1A2 | C1A3 | C1B1 | C1B2 | C1B3 | C1C1 | C1C2 | C1C3 |
| C2 | C2A1 | C2A2 | C2A3 | C2B1 | **C2B2** | **C2B3** | C2C1 | **C2C2** | **C2C3** |
| C3 | C3A1 | C3A2 | C3A3 | C3B1 | **C3B2** | **C3B3** | C3C1 | **C3C2** | **C3C3** |

Recessive of genotype is16 out of 81 populations. If there are two dominants in father, and two dominant in mother, the result would be one recessive and 80 dominants, Table (4).

Table (4):Two dominants in each of the parent.

| | A1 | A2 | A3 | B1 | B2 | B3 | C1 | C2 | C3 |
|---|---|---|---|---|---|---|---|---|---|
| A1 | A1A1 | A1A2 | A1A3 | A1B1 | A1B2 | A1B3 | A1C1 | A1C2 | A1C3 |
| A2 | A2A1 | A2A2 | A2A3 | A2B1 | A2B2 | A2B3 | A2C1 | A2C2 | A2C3 |
| A3 | A3A1 | A3A2 | A3A3 | A3B1 | A3B2 | A3B3 | A3C1 | A3C2 | A3C3 |
| B1 | B1A1 | B1A2 | B1A3 | B1B1 | B1B2 | B1B3 | B1C1 | B1C2 | B1C3 |
| B2 | B2A1 | B2A2 | B2A3 | B2B1 | B2B2 | B2B3 | B2C1 | B2C2 | B2C3 |
| B3 | B3A1 | B3A2 | B3A3 | B3B1 | B3B2 | B3B3 | B3C1 | B3C2 | B3C3 |
| C1 | C1A1 | C1A2 | C1A3 | C1B1 | C1B2 | C1B3 | C1C1 | C1C2 | C1C3 |
| C2 | C2A1 | C2A2 | C2A3 | C2B1 | C2B2 | C2B3 | C2C1 | C2C2 | C2C3 |
| C3 | C3A1 | C3A2 | C3A3 | C3B1 | C3B2 | C3B3 | C3C1 | C3C2 | **C3C3** |

Only one recessive is in 81 populations.

The number of dominants follows the equation:

$$D = (n_f . n_m)^2 - ((n_f - n_{fd}) . (n_m - n_{md}))^2 \quad (1)$$

Where D is the number of dominants.

$n_f$ is the number of alleles in father,
$n_m$ is the number of alleles in mother,
$n_{fd}$ is the number of dominants in father,
$n_{md}$ is the number of dominants in mother.

Maximizing equation (1) by differentiating and equal to zero, it follows that the larger the number of dominant genotypes is the smaller number of recessive ones.

If the mother has no dominant, then equation (1) will depend only on the crossing between the number of alleles in both parents and the dominant in the father.

Equation (1) is equivalent to Hardy-Weinberg equilibrium which is $p^2+2pq + q^2 =100\%$. Equating both equations, one can get:

$$(n_f . n_m)^2 = 100\% \quad (2)$$
$$D = p^2+2pq \quad (3)$$
$$((n_f - n_{fd}) . (n_m - n_{md}))^2 = q^2 \quad (4)$$

And therefore:

$$(p^2+2pq)/ q^2 = (n_{f.} n_m)^2 - ((n_f-n_{fd}) . (n_m-n_{md}))^2 / ((n_f-n_{fd}) . (n_m-n_{md}))^2 \quad (5)$$

If the number of dominant alleles in father or mother equals the number of traits, the dominator will be equal to zero, and therefore the whole population will be dominants.

Now consider a man and woman from the Amazon Forest married to each other; their dominant phenotypes are intact and undamaged due to the constant environment and the natural food which does not contain chemicals or pollutants. The product of the genotype population will all be dominant. However, if their first offspring is married to a tall and white man from Greenland, their offspring would probably be a mixture between dominant and recessive. But how can this offspring from Amazon get married to that offspring from Greenland without overcoming transportations obstacles and other related barriers. The question aroused that if no change in the environment and no change in social behavior, would genotypes have recessive traits. In other word, is there any evolution as per Darwinian theory if there is no change in the natural world. This is a simple way to prove that natural selection is mathematically impossible of producing evolutionary change, because there is no gene drift when Amazonians married each other.

To prove that Darwin theory can not be accepted, let us assume that all population are married to a father of pure recessive alleles, for the worst case scenario, then offspring would have recessive alleles.

### 1.16.2 Molecule Inheritance

By 1926, the search to determine the mechanism for genetic inheritance had reached the molecular level. Previous discoveries by Gregor Mendel, Thomas Hunt Morgan and Walter Sutton, and numerous other scientists had narrowed the search to the chromosomes located in the nucleus of most cells. But the question of what molecule was actually the genetic material had not been answered.

DNA and proteins are made of molecules. DNA is a long polymer of deoxyribonucleotides. The central concept of genetics involves the DNA-to-protein sequence involving transcription and translation. A gene does not directly control protein synthesis; instead, it passes its genetic information on to RNA, which is more directly involved in protein synthesis. DNA has a sequence of bases that is transcribed into a sequence of bases in mRNA. Every three bases is a codon that stands for a particular amino acid. The change of just one nucleotide causing a codon change can cause the wrong amino acid to be inserted in a polypeptide; this is a point mutation.

Since DNA is the genetic material, its structure and functions constitute the molecular basis of inheritance. Though there is no ambiguity that the genes are located on the DNA, it is difficult to literally define a gene in terms of DNA sequence. Although the structure of DNA showed how inheritance works, it was still not known how DNA influences the behavior of cells. Because the DNA molecule is able to replicate, genetic information can be passed from one cell generation to the next. DNA codes for the synthesis of proteins; this process also involves RNA. The *central concept of genetics* involves the DNA-to-protein sequence involving transcription and translation.

## 1.17 Inherited Traits

Inheritance means that a gene, chromosome or genome is transmitted from parents to child. For example, for the gene that determines eye colour, you may inherit a brown allele from your mother and a blue allele from your father. With the union of the egg and sperm, the individual begins life with a mixture (admixture) of Autosomal DNA from both parents. This admixture is not a pure 50%-50%. It will vary from child to child and is the determining factor for the differences among siblings. Identical twins share identical DNA where fraternal twins do not share identical DNA. The Y chromosome is only received by the son only from his father and both sons and daughters receive their mtDNA (mitochondria DNA) only from their mothers.

In the statistical world of DNA, on the average, we believe that each generation receives roughly half of the DNA of the generations before them. We know that each child absolutely receives 50% of the DNA of both parents, but how the grandparents DNA is divided up into that 50% that goes to each offspring differs. It may not be 50%. Let's use the 50% rule here, because it's all we have and it's what we've been working with forever.

In a normal autosomal, every generation provides to the current generation the following approximate % of DNA, Table (5).

| Generation | Average percentage of contributed DNA |
|---|---|
| GGGGG-Grandparents | 1.56% |
| GGGG-grandparents | 3.12% |
| Great-Great-grandparents | 6.25% |
| Great-grandparents | 12% |
| Grandparents | 25% |
| Parents | 50% |
| You | 100% |

1. Genetic traits are a difficult inheritance process.
2. The exact mixture of chromosome mixture at conception changes and does not follow a simple formula.
3. mtDNA DNA is different than the other autosomal DNA. It is not not a mixture from either parent, but supplied by only one parent.
4. The X-chromosome for the male is taken directly from his mother, his Y-chromosome is taken directly from his father. The X-chromosomes for a female are an admixture of her mother's X-chromosomes and her father's X-chromosome on average about 50% from each. This admixture is what makes siblings different. Identical twins share the same DNA admixture, but all the rules above still apply, Figure (27).

Figure (27): Inheritance of mtDNA of mothers

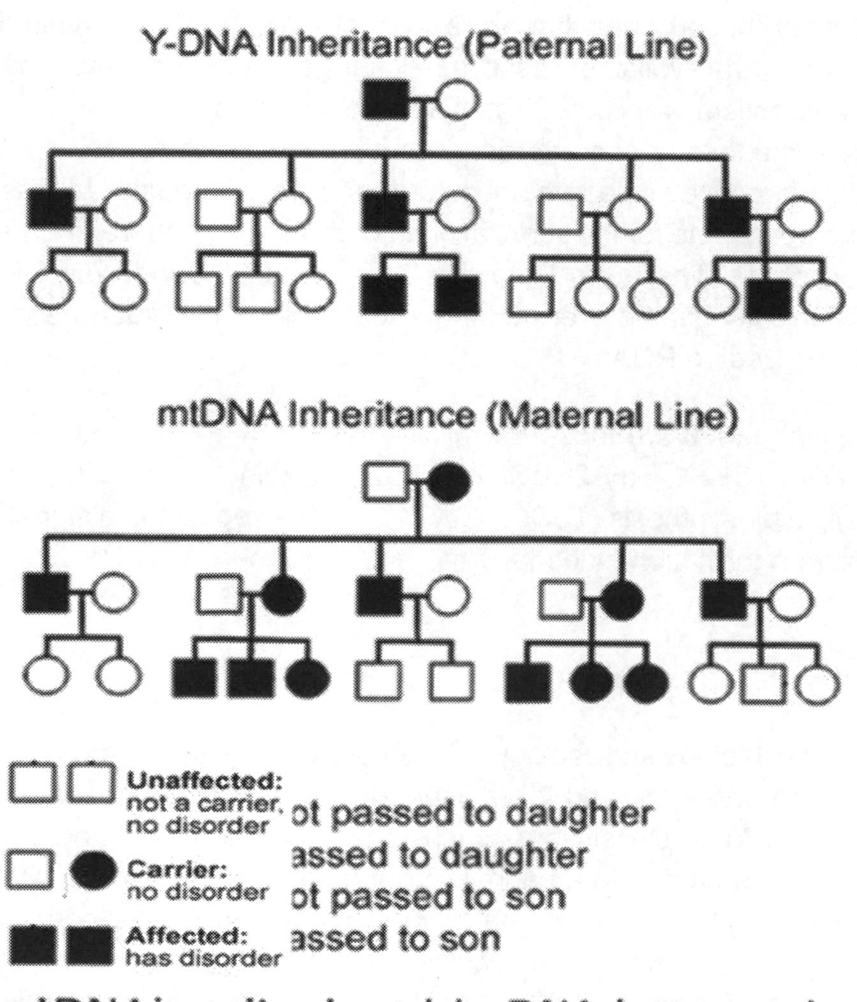

### 1.17.1 Acquired traits

Acquired traits are traits that occur after an organism is born. They are not passed down from one generation to the next. Acquired traits can be caused after birth by disease, injury, accident, deliberate modification, repeated use, disuse, or misuse, or other environmental

## 1.18 Essential Genes

Essential genes are those genes of an organism that are thought to be critical for its survival. However, being *essential* is highly dependent on the circumstances in which an organism lives. For instance, a gene required to digest starch is only essential if starch is the only source of energy. Recently, systematic attempts have been made to identify those genes that are absolutely required to maintain life, provided that all nutrients are available These essential genes encode proteins to maintain a central metabolism, replicate, translate into proteins, maintain a basic cellular structure, and mediate transport processes into and out of the cell. Most genes are not essential but convey selective advantage and increased fitness. Essential genes are those indispensable for the survival of an organism, and therefore are considered a foundation of life. There are Data Base of Essential Genes (DEG) which hosts records of currently available essential genomic elements, such as protein-coding genes and non-coding RNAs.

Some essential genes are those for mRNA splicing (PRP38, PRP24, PRP16), cell cycle and DNA (CKS1, RFA2, PSF2, RTS2), protein targeting (KAP92, HSP60), membrane lipid biosynthesis (TSC11, GWT1, GP110), regulation o signal transduction (FRQ1), enzyme formation, and many more.

## 1.19 Genetic Polymorphism

Polymorphism in biology and zoology is the occurrence of two or more clearly different morphs or forms, also referred to as alternative phenotypes, in the population of a species. In order to be classified as such, morphs must occupy the same habitat at the same time and belong to a panmictic population (A panmictic population is one where all individuals are potential partners. This assumes that there are no mating restrictions, neither genetic nor behavioral, upon the population, and that therefore all recombination is possible. The Wahlund effect assumes that the overall population is panmictic, in other words, one with random mating).

There are three types of genetic polymorphism:

- Genetic polymorphism - where the phenotype of each individual is genetically determined.
- A conditional development strategy, where the phenotype of each individual is set by environmental cues.
- A mixed development strategy, where the phenotype is randomly assigned during development.

The genetic alteration can happen in many forms. The most common is the one that can change the building block of the DNA and RNA. This type of alteration is called Single Nucleotide Polymorphism (SNP). SNP can change the nucleotide T to G or C, Figure (28).

Some of the genetic SNP have an effect on human health. For example, the can utilized to trace the inheritance of disease genes in the family. Disease genes are due to the changes in the nucleotides altered by SNPs, as they occur once in every 300 nucleotide on average, It is estimated that the human genome has about 10 million SNPs.

Figure (28): Genetic variation due to SNP

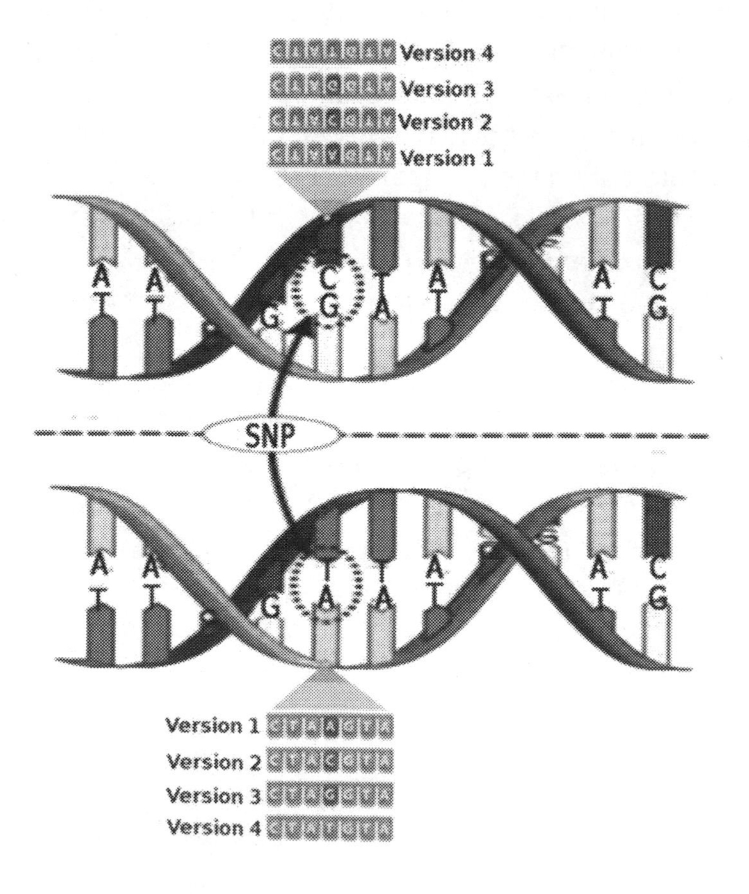

# CHAPTER 2

# EPIGENETICS

## 2.1 Introduction

In the science of genetics, epigenetics is the study of cellular and physiological phenotypic trait variations that result from external or evvironmental factors that switch genes on and off and affect how cells express genes. The term epigenetics refers to heritable changes in gene expression (active versus inactive genes) that does not involve changes to the underlying DNA sequence; a change in phenotype without a change in genotype. This in turn affects how cells read the genes. Epigenetic change is a regular and natural occurrence but can also be subjective by several factors including age, the environment/lifestyle, and disease state. Behavioral epigenetics is the field of study examining the role of epigenetics in shaping animal (including human) behaviour It is an experimental science that seeks to explain how nurture shapes nature, where nature refers to biological heredity and nurture refers to virtually everything that occurs during the life-span (e.g., social-experience, diet and nutrition, and exposure to toxins). Behavioral epigenetics attempts to provide a framework for understanding how the expression of genes is influenced by experiences and the environment to produce individual differences in bhaviour, cognition personality, and mental health.

At least three systems jointly with DNA methylation, histone modification and non-coding RNA (ncRNA) -associated gene silencing are currently considered to initiate and sustain epigeneticvariation. New and ongoing research is always revealing the role of epigenetics in a variety of human disorders and fatal diseases. The modification of chromatin (the tightly packed complex of proteins and genomic DNA) is responsible for epigenetic regulation of gene expression. Like genetic changes, epigenetic changes are preserved when a cell divides. A cell's epigenome is the overall epigenetic state of a cell.

## 2.2 Molecular Basis of Epigenetics DNA Modification and Histone Modification

The term also refers to the changes themselves: functionally relevant changes to the genome that do not involve a change in the nucleotide sequence. Examples of mechanisms that produce such changes are DNA methylation and histone modification. DNA methylation, and small non-coding RNAs. Diet, pollution, infections, and other environmental factors have great effects on epigenetic modifications and trigger susceptibility to diseases. Despite a growing body of literature addressing the role of the environment on gene expression, very little is established about the epigenetic pathways involved in the modulation of inflammatory and anti-inflammatory genes. This review summarizes the current knowledge about epigenetic control mechanisms during the inflammatory response.

Chromatin is the complex of chromosomal DNA associated with proteins in the nucleus. DNA in chromatin is packaged around histone proteins, in units referred to as nucleosomes. A nucleosome has 147 base pairs of DNA associated with an octomeric core of histone proteins, which consists of 2 H3-H4 histone dimers surrounded by 2 H2A-H2B dimers. N-terminal histone tails extend beyond from nucleosomes into the nuclear lumen. H1 histone associates with the linker DNA located between the nucleosomes. Nucleosome spacing controls and determines chromatin structure, which can be broadly divided into heterochromatin and euchromatin. Chromatin structure and gene accessibility to transcriptional machinery are regulated by modifications to both DNA and histone tails, Figure (29).

Figure (29): Structure of DNA

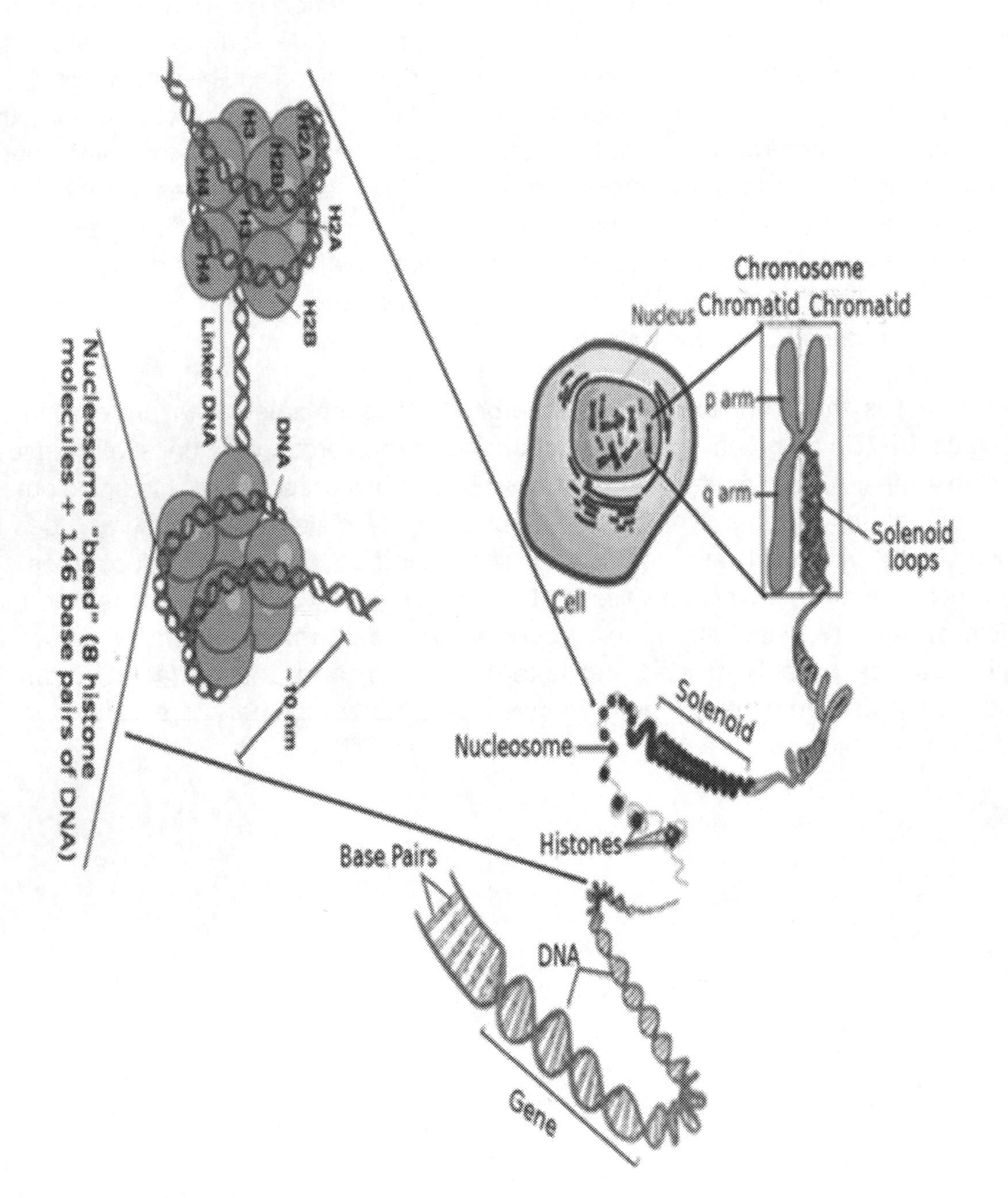


## 2.3 Histone Modification and Effect

Histone methylation is a process by which methyl groups are transferred to amino acids of histone proteins that make up nucleosomes, which the DNA double helix wraps around to form chromosomes. Methylation of histones can either increase or decrease transcription of genes, depending on which amino acids in the histones are methylated, and how many methyl groups are attached. Methylation events that weaken chemical attractions between histone tails and DNA increase transcription, because they enable the DNA to uncoil from nucleosomes so that transcription factor proteins and RNA polymerase can access the DNA. This process is critical for the regulation of gene expression that allows different cells to express different genes.

### 2.3.1 Epigenetic Modifications and Mechanisms

Covalent modifications of either DNA (e.g. cytosine methylation and hydroxymethylation) or of histone proteins (e.g. lysine acetylation, lysine and arginine methylation, serine and threonine phosphorylation, and lysine ubiquitination and sumoylation) play central roles in many types of epigenetic inheritance. Therefore, the word "epigenetics" is sometimes used as a synonym for these processes. However, this can be misleading. Chromatin remodeling is not always inherited, and not all epigenetic inheritance involves chromatin remodeling.

DNA associates with histone proteins to form chromatin.

Because the phenotype of a cell or individual is affected by which of its genes are transcribed, heritable transcription can give rise to epigenetic effects. There are several layers of regulation of gene expression. One way that genes are regulated is through the remodeling of chromatin. Chromatin is the complex of DNA and the histone proteins with which it associates. If the way that DNA is wrapped around the histones changes, gene expression can change as well. Chromatin remodeling is accomplished through two main mechanisms:

1. The first way is post translational modification of the amino acids that make up histone proteins. Histone proteins are made up of long chains of amino acids. If the amino acids that are in the chain are changed, the shape of the histone might be modified. DNA is not completely unwound during replication. It is possible, then, that the modified histones may be carried into each new copy of the DNA. Once there, these histones may act as templates, initiating the surrounding new histones to be shaped in the new manner. By altering the shape of the histones around them, these modified histones would ensure that a lineage-specific transcription program is maintained after cell division.
2. The second way is the addition of methyl groups to the DNA, mostly at CpG sites, to convert cystosine to 5-methyl cystosine. 5-Methylcytosine performs much like a regular cytosine, pairing with a guanine in double-stranded DNA. However, some areas of the genome are methylated more heavily than others, and highly methylated areas tend to be less transcriptionally active, through a mechanism not fully understood. Methylation of cytosines can also persist from

the germ line of one of the parents into the zygote, marking the chromosome as being inherited from one parent or the other (genetic imprinting).

Mechanisms of heritability of histone state are not well understood; however, much is known about the mechanism of heritability of DNA methylation state during cell division and differentiation. Heritability of methylation state depends on certain enzymes (such as DNMT1) that have a higher affinity for 5-methylcytosine than for cytosine. If this enzyme reaches a "hemimethylated" portion of DNA (where 5-methylcytosine is in only one of the two DNA strands) the enzyme will methylate the other half, Figure (30).

Figure (30): Effect of methylation on gene's activity

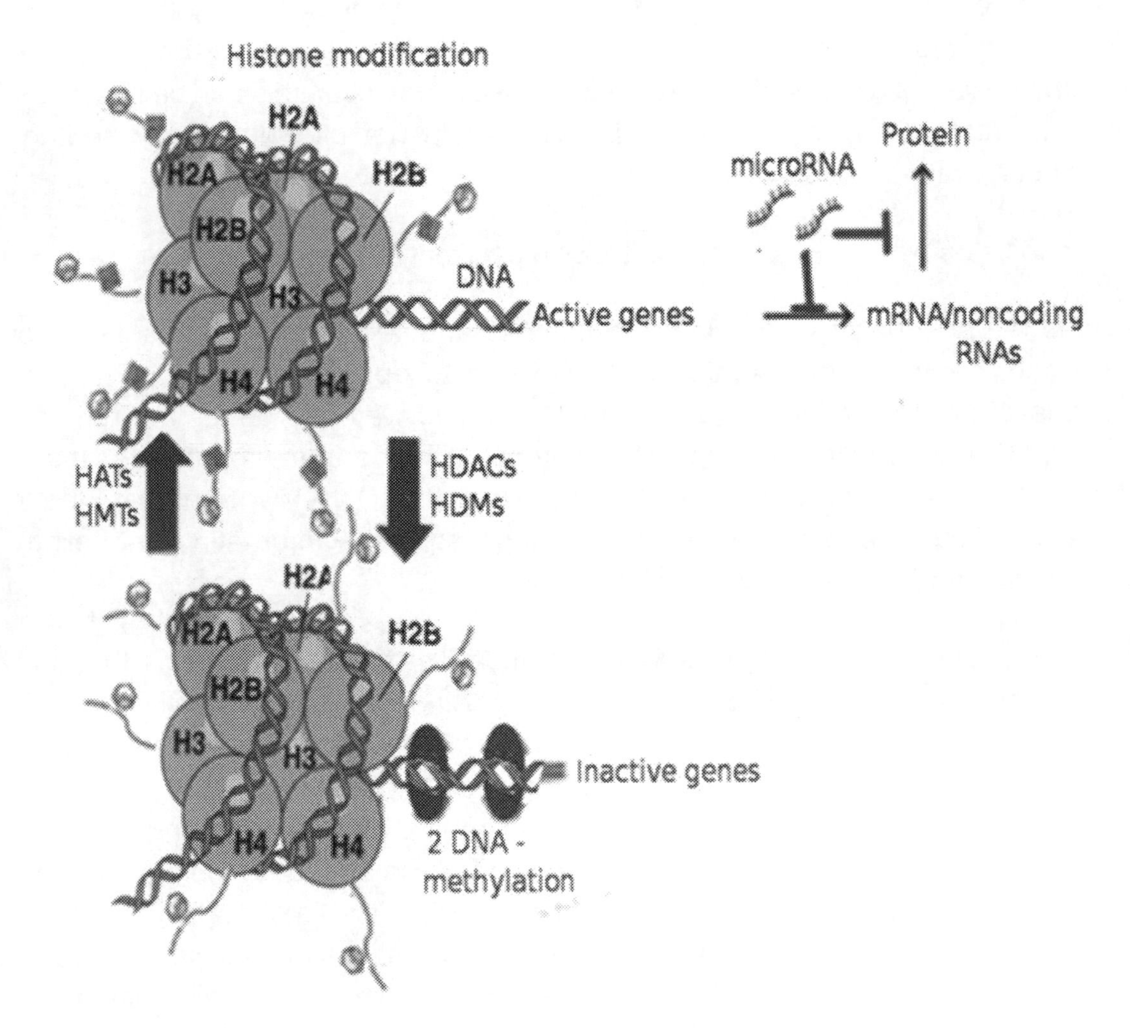


## 2.3.2 Acetylation

Histones are the proteins closely associated with DNA molecules and they are the chief protein components of chromatin as clarified earlier. Acetylation of the lysine residues at the N terminus of histone proteins removes positive charges, thereby reducing the affinity between histones and DNA. This makes RNA polymerase and transcription factors easier to access the promoter region. Therefore, in most cases, histone acetylation improves transcription while histone deacetylation represses transcription. Acetylation or ethanoylation can be achieved by the addition of acetyl group COCH3 to the histone as ethanol has two atoms of carbons.

Acetylation brings in a negative charge, acting to neutralize the positive charge on the histones and decreases the interaction of the N termini of histones with the negatively charged phosphate groups of DNA. As a consequence, the condensed chromatin is transformed into an extra relaxed structure which is associated with greater levels of gene transcription. This relaxation can be reversed by HDAC (histone deacetylate) activity. Relaxed, transcriptionally active DNA is referred to as euchromatin. More condensed (tightly packed) DNA is referred to as heterochromatin. Figure (24) shows acetylation and deacetylation of histone lysine. The mechanism for acetylation and deacetylation takes place on the NH3+ groups of Lysine amino acid residues. These residues are located on the tails of histones that make up the nucleosome of packaged dsDNA. The process is aided by factors known as Histone Acetyltransferases (HATs). HAT molecules facilitate the transfer of an acetyl group from a molecule of Acetyl Coenzyme-A (Acetyl-CoA) to the NH3+ group on Lysine. When a Lysine is deacetylated, factors known as Histone Deacetylases (HDACs) catalyze the removal of the acetyl group with a molecule of H2O. HAT molecules facilitate the transfer of an acetyl group from a molecule of Acetyl Coenzyme-A (Acetyl-CoA) to the NH3+ group on Lysine. When a Lysine is deacetylated, factors known as Histone Deacetylases (HDACs) catalyze the removal of the acetyl group with a molecule of H2O. Acetylation has the effect of changing the overall charge of the histone tail from positive to neutral. Nucleosome formation is dependent on the positive charges of the H4 histones and the negative charge on the surface of H2A histone fold domains. Acetylation of the histone tails disrupts this association, leading to weaker binding of the nucleosomal components.[1] By doing this, the DNA is more accessible and leads to more transcription factors being able to reach the DNA. Thus, acetylation of histones is known to increase the expression of genes through transcription activation. Deacetylation performed by HDAC molecules has the opposite effect. By deacetylating the histone tails, the DNA becomes more tightly wrapped around the histone cores, making it harder for transcription factors to bind to the DNA. This leads to decreased levels of gene expression and is known as gene silencing.

Activation and inactivation of gene transcription can not be completed without the linker histone H1 which represents the pillar of the chromatin. Studies have demonstrated that histones H1, H2A, H2B and H3 and H4 are effective mediators of transfection. Electron microscopy and biochemical studies have established that the bulk of the chromatin DNA is compacted into repeating structural units called nucleosomes. Each nucleosome contains about 200 base pairs (bp) of DNA associated with a histone octamer core consisting of two molecules each of histones H2A, H2B, H3, and H4. Positions and types of activation are shown in Figure (31).

Figure (31): Activation of gene's transcription by histons

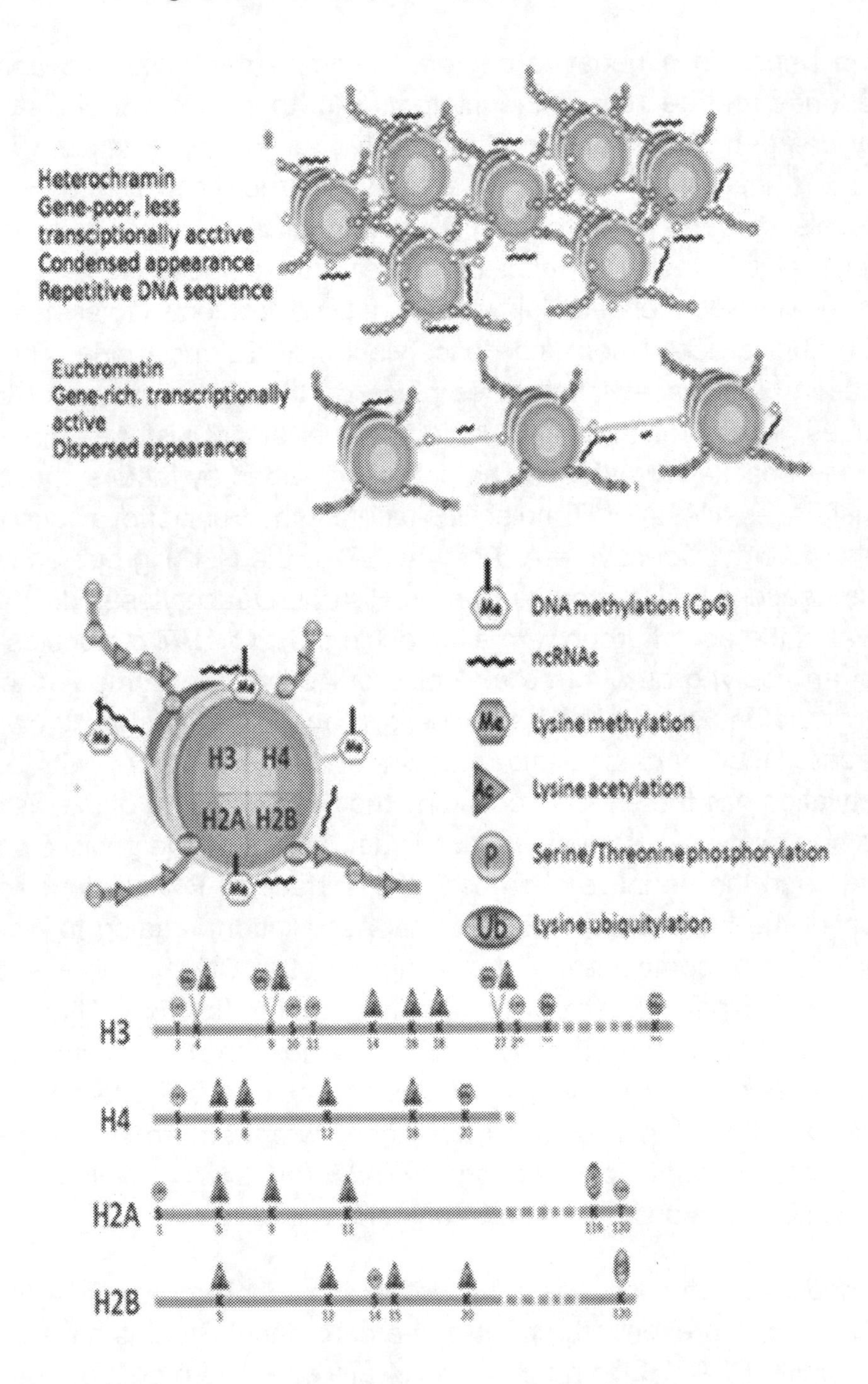


Acetylated histones, the octomeric protein cores of nucleosomes, represent a type of epigenetic marker within chromatin. Studies have shown that one modification has the tendency to influence whether another modification will take place. Modifications of histones can not only cause secondary structural changes at their specific points, but can cause many structural changes in distant locations which inevitably affect function. As the chromosome is replicated, the modifications that exist on the parental chromosomes are handed down to daughter chromosomes. The modifications, as part of their function, can recruit enzymes for their particular function and can contribute to the continuation of modifications and their effects after replication has taken place. [1] It has been shown that, even past one replication, expression of genes may still be affected many cell generations later. A study showed that, upon inhibition of HDAC enzymes by Trichostatin A, genes inserted next to centric heterochromatin showed increased expression. Many cell generations later, in the absence of the inhibitor, the increased gene expression was still expressed, showing modifications can be carried through many replication processes such as mitosis and meiosis, Figure (32).

Figure (32): Condensed and relaxed DNA due to acetylation and deacetylation

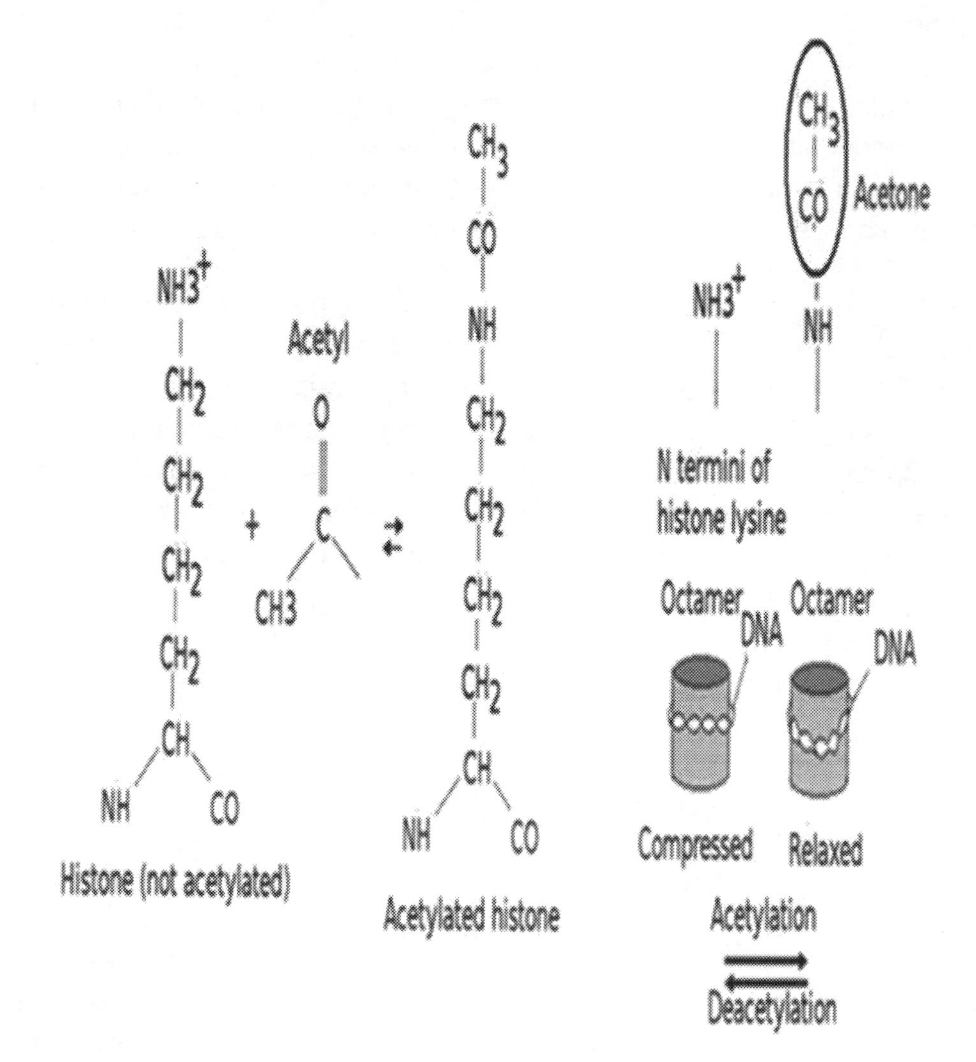

Greater level of gene transcription can elongate the life span and can be regulated to fight diseases.

Histone acetylation is catalyzed by histone acetyltransferases (HATs) and histone deacetylation is catalyzed by histone deacetylases (HDs). Several different forms of HATs and HDs have been identified. Among them, CBP/p300 is probably the most available, and can interact with numerous transcription regulators. The unusual properties of p300/CBP is one of the most important enzymes in the HAT family, and can be targeted for developing new anti-cancer drugs.

### 2.3.3 Effect of Epigenetic and Genetic Alteration

A variety of epigenetic mechanisms can be perturbed in different types of cancer. Epigenetic alterations of DNA repair genes or cell cycle control genes are very frequent in sporadic (non-germ line) cancers, being significantly more common than germ line (familial) mutations in these sporadic cancers. Epigenetic alterations are important in cellular transformation to cancer, and their manipulation holds great promise for cancer prevention, detection, and therapy. Several medications which have epigenetic impact are used in several of these diseases. These aspects of epigenetics are addressed in cancer epigenetics.

Chromatin remodeling is an epigenetic phenomenon. The three pillars of epigenetic regulation are DNA methylation, histone modifications and non-coding RNA species. Alterations of these epigenetic mechanisms affect the vast majority of nuclear processes including gene transcription and silencing, DNA replication and repair, cell cycle progression, and telomere and centromere structure and function. One of our research goals is to understand how epigenetic defects contribute to the pathophysiology of aging and aging-related diseases, especially cancer. Epigenetic changes may instigate aging and cancer phenotypes, or prime cells in such a way as to make them more susceptible to subsequent genetic or epigenetic alterations. The accumulation of further genetic or epigenetic changes over time would promote the progression of aging and cancer phenotypes.

The importance of histones and chromatin structure (histone octamer plus DNA are called nucleosome) in the regulation of eukaryotic gene transcription has become much more widely accepted over the past few years. It has been clear for a decade that histones and chromatin structure contribute to the regulation of transcription. Latest studies have led to the striking observation that several protein complexes engaged in transcription regulation can function, at least in part, by modifying histones or altering chromatin structure. While it is clear that many of these protein complexes have functions in addition to chromatin modification, they show the importance of chromatin structure as a part of transcription regulation mechanisms, Figure (33).

http://eurheartj.oxfordjournals.org/content/30/3/266

Figure (33): Remodeling of chromatin for transcription

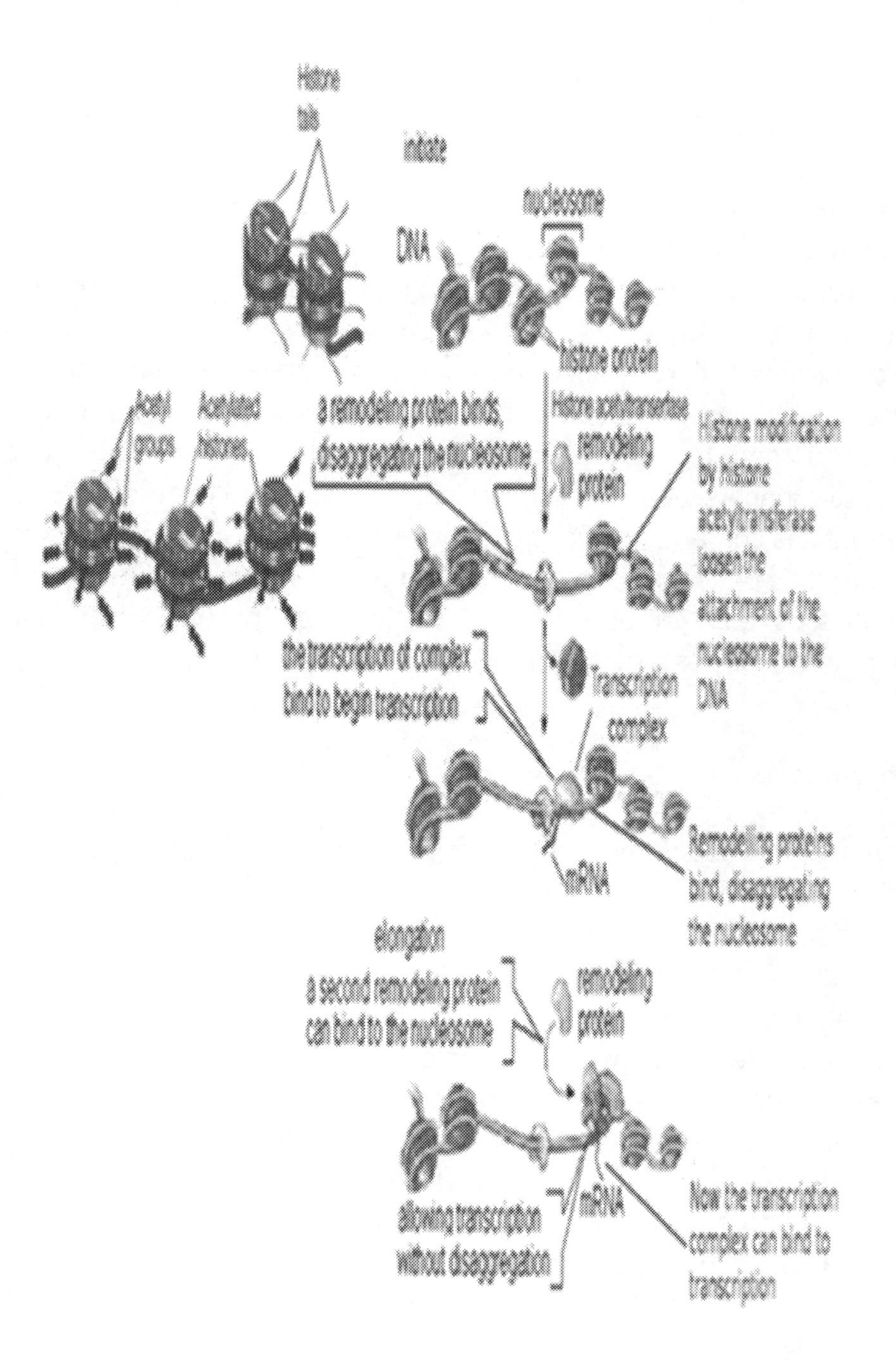

Activation and inactivation of gene transcription can not be completed without the linker histone H1 which represents the pillar of the chromatin. Studies have proved that histones H1, H2A, H2B and H3 and H4 are effective mediators of transfection. Electron microscopy and biochemical studies have established that the bulk of the chromatin DNA is compacted into repeating structural units called nucleosomes. Each nucleosome contains about 200 base pairs (bp) of DNA associated with a histone octamer core consisting of two molecules each of histones H2A, H2B, H3, and H4.

The basic unit of chromatin, the nucleosome, consists of DNA that wraps around an octamer of histones. The histone tails protrude from the nucleosome core and are subjected to different post-translational modifications. Nucleosomal DNA can also be methylated by DNA methyltransferase activities. The types of post-translational modifications of histones and the degree of DNA methylation influence a specific chromatin structure. Two morphologically distinct types of chromatin can be distinguished, namely euchromatin and heterochromatin. Euchromatin is linked with gene-rich and transcriptionally active domains of dispersed appearance characterized biochemically by hyperacetylation of histones and hypomethylation of both histones and DNA. In contrast, highly condensed heterochromatin is linked to gene-poor and transcriptionally inactive domains such as centromeres and telomeres. Biochemically, heterochromatin is characterized by hypoacetylation of histones, and hypermethylation of histones and DNA. In addition, non-coding RNAs have been recently recognized as players in chromatin remodeling, primarily involved in heterochromatin formation and transcriptional or post-transcriptional gene silencing.

### 2.3.4 Methylation

Methylation of DNA is a common method of gene silencing. DNA is typically methylated by methyltransferase enzymes on cytosine nucleotides in a CpG dinucleotide sequence (also called “CpG islands” when densely clustered). Analysis of the pattern of methylation in a given region of DNA (which can be a promoter) can be achieved through a method called bisulfite mapping. Methylated cytosine residues are unchanged by the treatment, whereas unmethylated ones are changed to uracil. The differences are analyzed by DNA sequencing or by methods developed to quantify SNPs, measuring the relative amounts of C/T at the CG dinucleotide. Abnormal methylation patterns are thought to be involved in oncogenesis. Histone acetylation is also an important process in transcription. Histone acetyltransferase enzymes (HATs) such as CREB-binding protein also dissociate the DNA from the histone complex, allowing transcription to proceed. Often, DNA methylation and histone deacetylation work together in gene silencing. The combination of the two seems to be a signal for DNA to be packed more densely, lowering gene expression,

In general, but not always, DNA methylation is associated with loss of gene expression. One theory on the evolution of DNA methylation is that it evolved as a host defence mechanism to silence foreign DNA such as viral sequences, replicated transposable elements and other repetitive sequences.

In normal cells, promoter of some genes such as tumor suppressor protein, DNA repair proteins is unmethylated and accessible to binding to the transcription factors (TF) allowing transcription. But, in many cancers these genes are methylated by DNA methyltransferase 1 and therefore bound by the methyl-CpG binding proteins (MBD) and histone deacetylase (HDAC). Thus the methylated promoter is not accessible to binding to the transcription factors and inactive. In tumor tissues and biological fluids such as serum and urine, the methylated DNA is measured by various methods for the development of diagnostic and prognostic tools for the cancer. HAT indicates histone acetyltransferase; RNA pol II, RNA polymerase II; and HMT, histone methyltransferase, Figure (34).

Figure (34): Methylation of cytosine

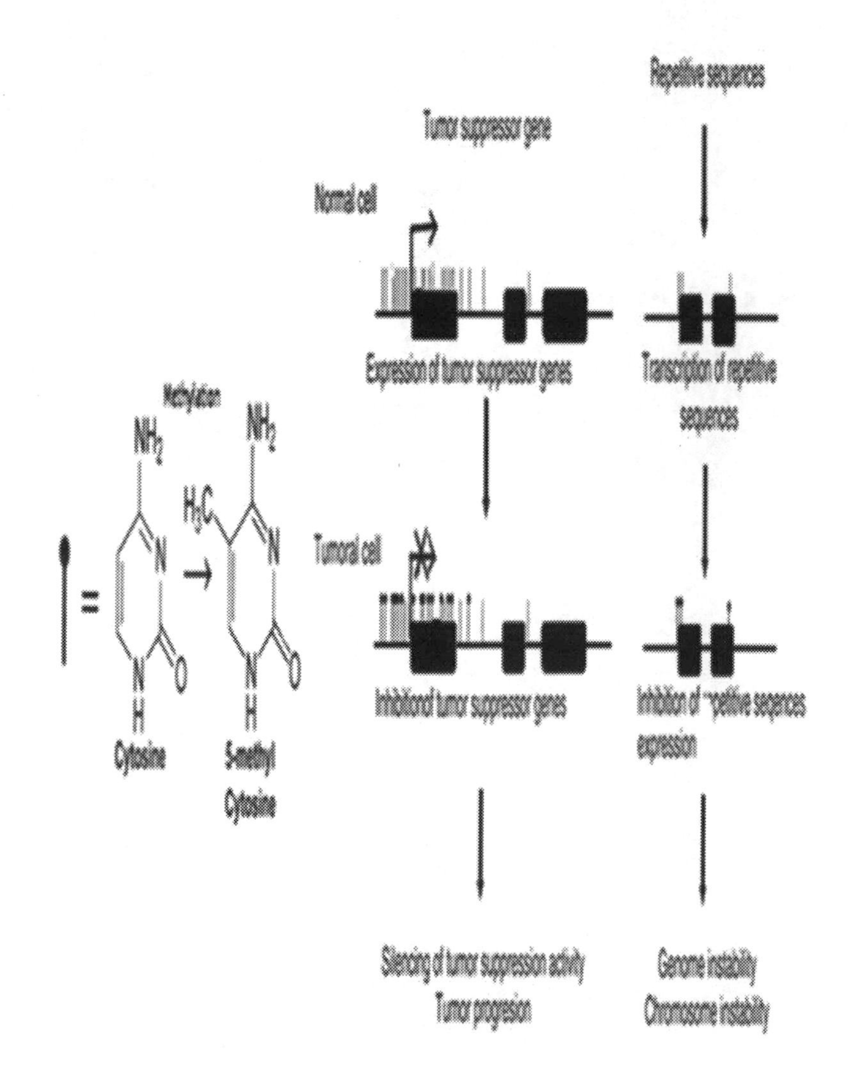

Epigenetic alterations, mainly in DNA methylation, modification of histones, and expression of non-coding RNAs are recognized as mechanisms contributing to malignancy. One of the aims of epigenetic alterations is the characterization of epigenetic mechanisms that regulate chromatin structure and the impact that epigenetic alterations have on chromosome stability, tumorigenesis and radiosensitivity.

Methods of epigenetic alteration are: acetylation, methylation, phosphorylation, ubiquitylation, sumoylation, ADP ribosylation, deimination, biotinylation, butyrylation, N-formylation, and proline isomerization. Figure (35) shows the effect of methylation and acetylation on transcription.

Figure (35): Effect of methylation and acetylation on transcription

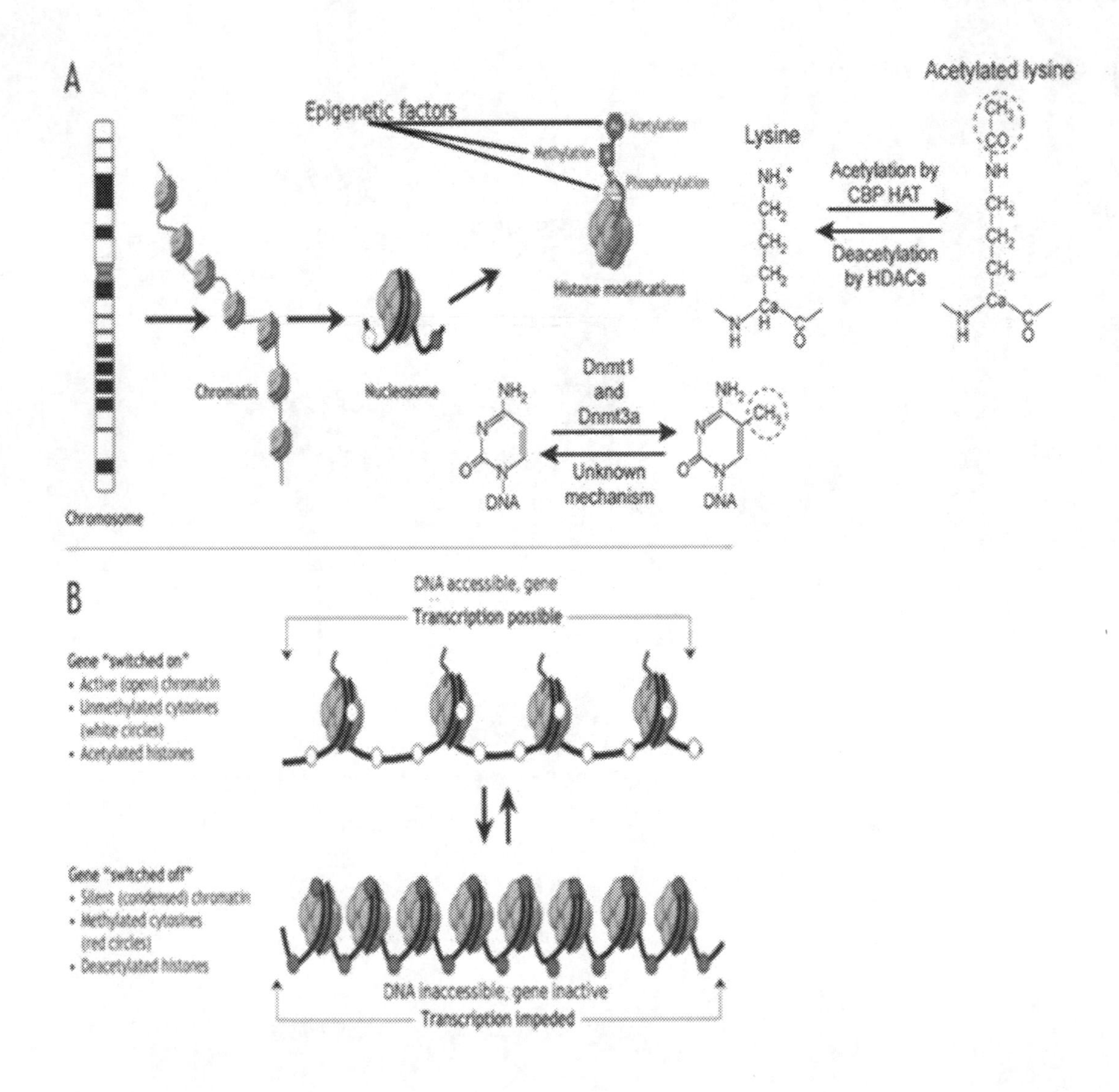

## 2.3.5 Histone Protein and Non Histone Protein

In chromatin, those proteins which remain after the histones have been removed, are classified as non-histone proteins. Both are proteins, both provide structure to DNA, and both are components of chromatin. Their chief difference is in the structure they provide. Histone proteins are the spools about which DNA winds, whereas nonhistone proteins provide the scaffolding structure. Another way to think of the difference is that nonhistone proteins are those proteins remaining after all histones have been removed from chromatin. Protein methylation involves in many important biological processes including transcription activity, signal transduction and regulation of gene expression. Most of recent studies focus on lysine methylation of histones due to its critical roles of regulation in transcriptional repression or activation. Histones possess highly conserved sequences and are homologous in most species. However, there is much less sequence conservation within non-histone proteins. Thus, the characteristics of lysine methylation sites may be quite different between histones and non-histone proteins, Figure (36).

Figure (36): Histone protein and non-histone protein

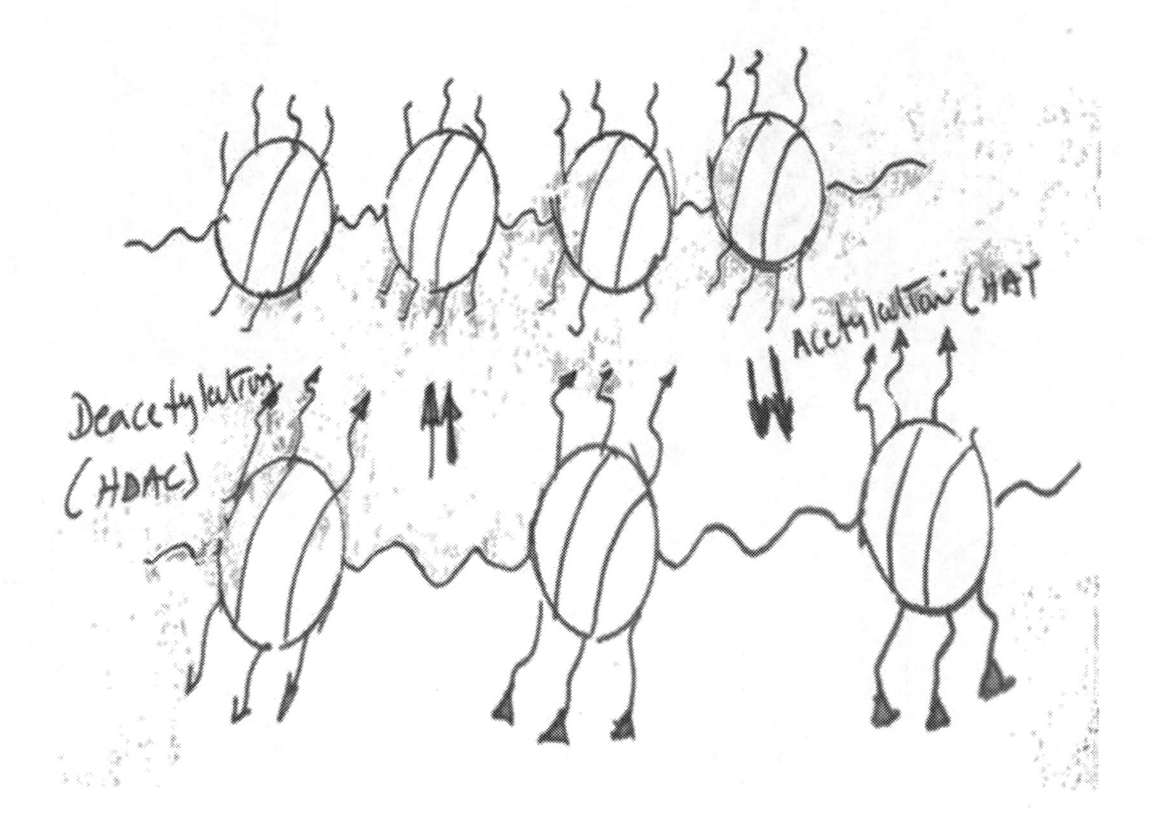

### 2.3.6 Mechanisms of DNA Methylation and Demethylation, DNA Replication

DNA methylation typically acts to repress gene transcription. DNA methylation is essential for normal development and is associated with a number of key processes including genomic imprinting, X-chromosome inactivation, repression of repetitive elements, aging and carcinogenesis.

DNA methylation patterns are established during embryonic development. During early development, methylation patterns are initially established by the de novo DNA methyl transferases DNMT3A and DNMT3B. When DNA replication and cell division occur, these methyl marks are maintained in daughter cells by the maintenance methyltransferase, DNMT1, which has a preference for hemi-methylated DNA. If DNMT1 is inhibited or absent when the cell divides, the newly synthesized strand of DNA will not be methylated and successive rounds of cell division will result in passive demethylation. By contrast, active demethylation can occur through the enzymatic replacement of 5-methylcytosine (5meC) with Cytosine, Figure (37).

Figure (37): Effect of methylation on DNA replication

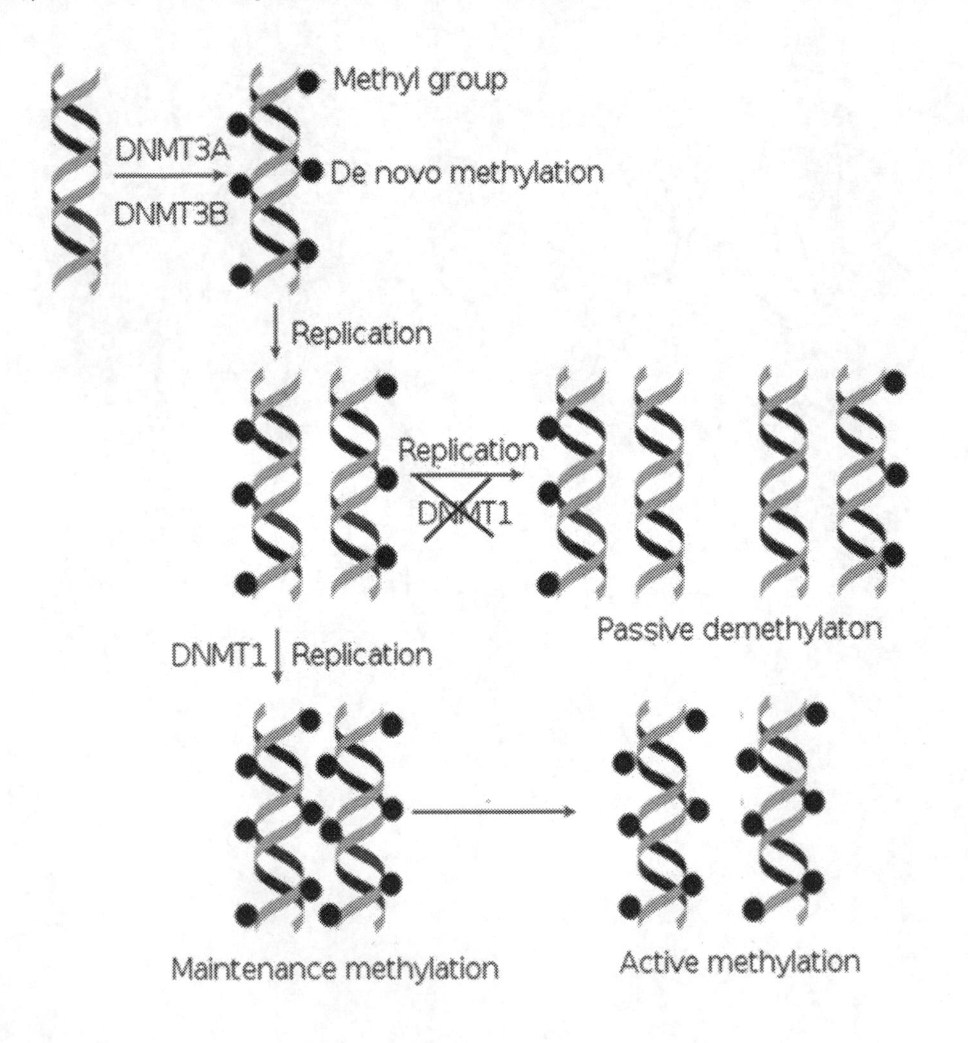

### 2.3.7 DNA Methylation and Cancer

Epigenetic changes such as DNA methylation act to regulate gene expression in normal mammalian development. However, promoter hypermethylation also plays a major role in cancer through transcriptional silencing of critical growth regulators such as tumor suppressor genes. Cancer-associated DNA hypomethylation is as prevalent as cancer-linked hypermethylation, but these two types of epigenetic abnormalities usually seem to affect different DNA sequences. Too much methyl (overmethylation) results in a unique set of symptoms. While undermethylation is generally considered less problematic than overmethylation. Much more of the genome is generally subject to undermethylation rather than overmethylation. Genomic hypermethylation in cancer has been observed most often in CpG islands in gene regions. In contrast, very frequent hypomethylation is seen in both highly and moderately repeated DNA sequences in cancer, including heterochromatic DNA repeats, dispersed retrotransposons, and endogenous retroviral elements. Other chromatin modifications, such as histone deacetylation and chromatin-binding proteins, affect local chromatin structure and, in concert with DNA methylation, regulate gene transcription. One of the best-known lesions of the malignant cell is the transcriptional repression of tumor-suppressor genes by promoter CpG island hypermethylation The DNA methylation inhibitors azacitidine and decitabine can induce functional re-expression of aberrantly silenced genes in cancer, causing growth arrest and apoptosis in tumor cells. These agents, along with inhibitors of histone deacetylation, have shown clinical activity in the treatment of certain hematologic malignancies where gene hypermethylation occurs. This review examines alteration in DNA methylation in cancer, effects on gene expression, and implications for the use of hypomethylating agents in the treatment of cancer, Figure (38).

Figure (38): Hypermethylation and hypomethylation

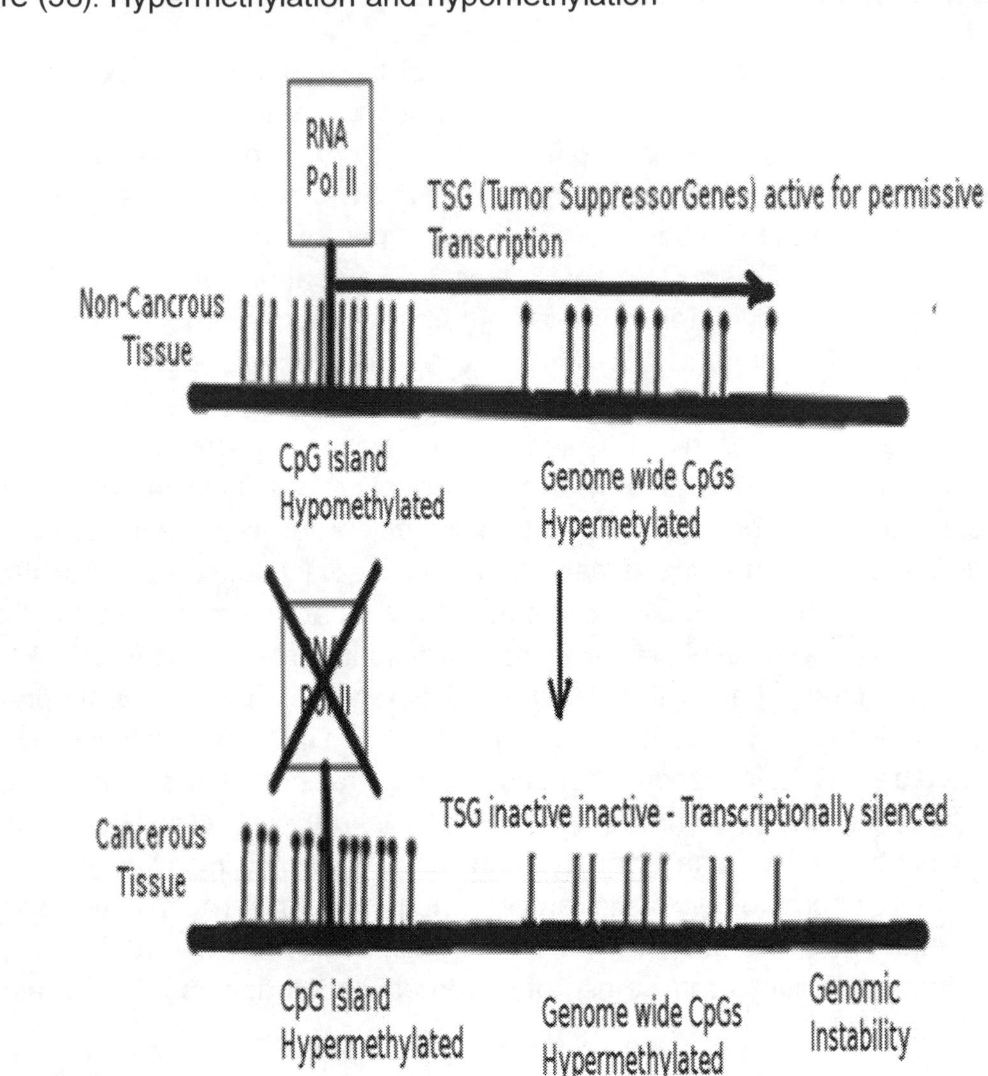


### 2.3.8 Chromatin-Binding Proteins

- Within chromosomes, DNA is held in complexes with structural proteins. These proteins organize the DNA into a compact structure called chromatin. DNA methylation is a most important epigenetic modification in the genomes of higher eukaryotes. In vertebrates, DNA methylation occurs predominantly on the CpG dinucleotide, and approximately 60% to 90% of these dinucleotides are modified. Sites of DNA methylation are occupied by various proteins, including methyl-CpG binding domain (MBD) proteins which recruit the enzymatic machinery to establish silent chromatin; MBD is a protein that in humans is encoded by the *MBD* gene. The protein encoded by MBD (there are two MBDs; MBD1 and MBD2 proteins) joins to methylated sequences in DNA, and thereby influences transcription. It binds to a range of methylated sequences, and appears to mediate repression of gene expression. Each of these proteins is capable of binding in particular to methylated DNA, especially each of a methyl-CpG Mutations in the MBD family member MeCP2 are the cause of Rett syndrome, a severe neurodevelopmental disorder, whereas other MBDs are known to bind sites of hypermethylation in human cancer cell lines. Here, we examine the advances in our understanding of the function of DNA methylation, DNA methyltransferases, and methyl-CpG binding proteins in vertebrate embryonic development. MBDs function in transcriptional repression and long-range interactions in chromatin and also appear to play a role in genomic stability, neural signaling, and transcriptional activation. DNA methylation makes an essential and versatile epigenetic contribution to genome integrity and function, Figure (39).

Figure (39): Epigenetic modification using DNA methylation

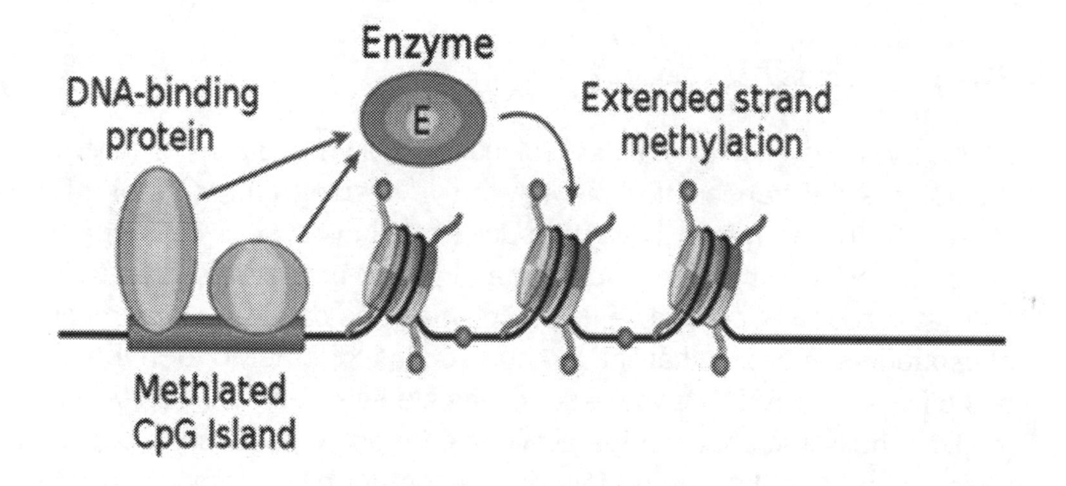


- AGO and PiWi regulatory Proteins

The Argonaute protein family plays a central role in RNA silencing processes, as essential catalytic components of the RNA-induced silencing complex (RISC). RISC complex is responsible for the gene silencing phenomenon known as RNA interference (RNAi). Argonaute proteins bind different classes of small non-coding RNAs, including microRNAs (miRNAs), small interfering RNAs (siRNAs) and Piwi-interacting RNAs (piRNAs). Small RNAs guide Argonaute proteins to their specific targets through sequence complementarity (base pairing), which then leads to mRNA cleavage or translation inhibition, Figure (40).

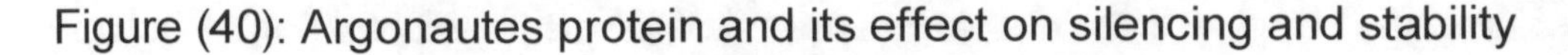

Figure (40): Argonautes protein and its effect on silencing and stability

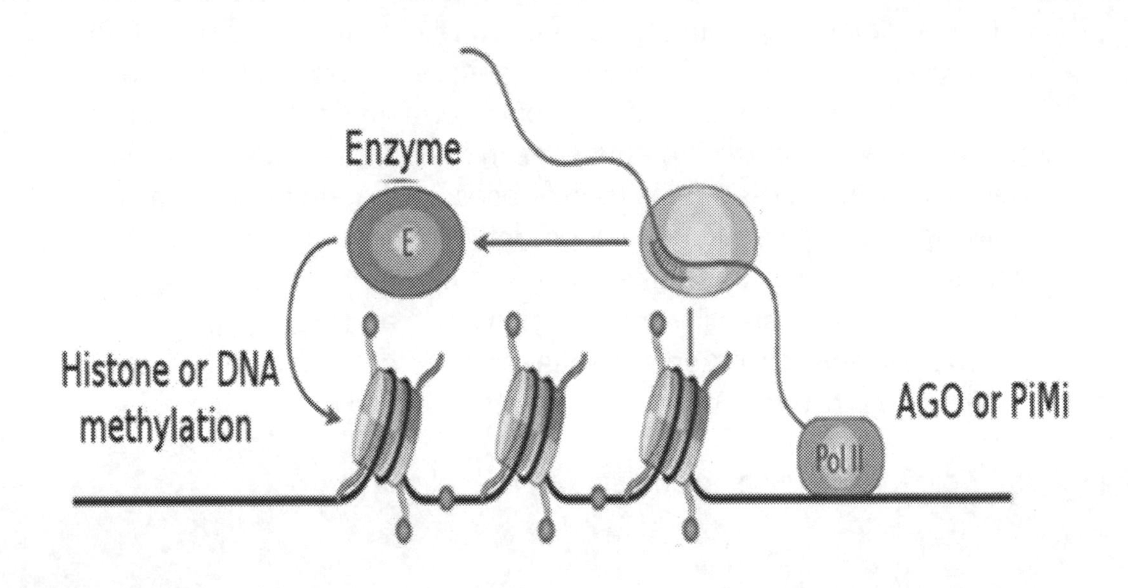


- RNA polymerase II (Pol ii)

RNA polymerase II (RNAP II and Pol II) is an enzyme found in eukaryotic cells. It catalyzes the transcription of DNA to synthesize precursors of mRNA and most snRNA and microRNA. The nuclei of all eukaryotic cells contain three different RNA Polymerases, designated I, II and III. Like the DNA Polymerase that catalyzes DNA replication, RNA Polymerases catalyze the formation of the phosphodiester bonds that link the nucleotides together to form the bsck nonr of the DNA. The RNA Polymerase moves stepwise along the DNA, unwinding the DNA helix just ahead of the active site for polymerization to expose a new region of the template strand for complementary base-pairing. Pol ii helps for DNA replications through substrates of the triphosphate ATP. Each eukaryotic RNA polymerase catalyzes transcription of genes encoding various classes of RNA. RNA polymerase catalyzes transcription of all protein-coding genes like mRNAs. RNA Polymerase-ii also produces four snRNAs (small nuclear RNAs) that take part in RNA splicing, Figure (41).

Figure (41): Eukaryotic RNA polymerase for catalyzing mRNA

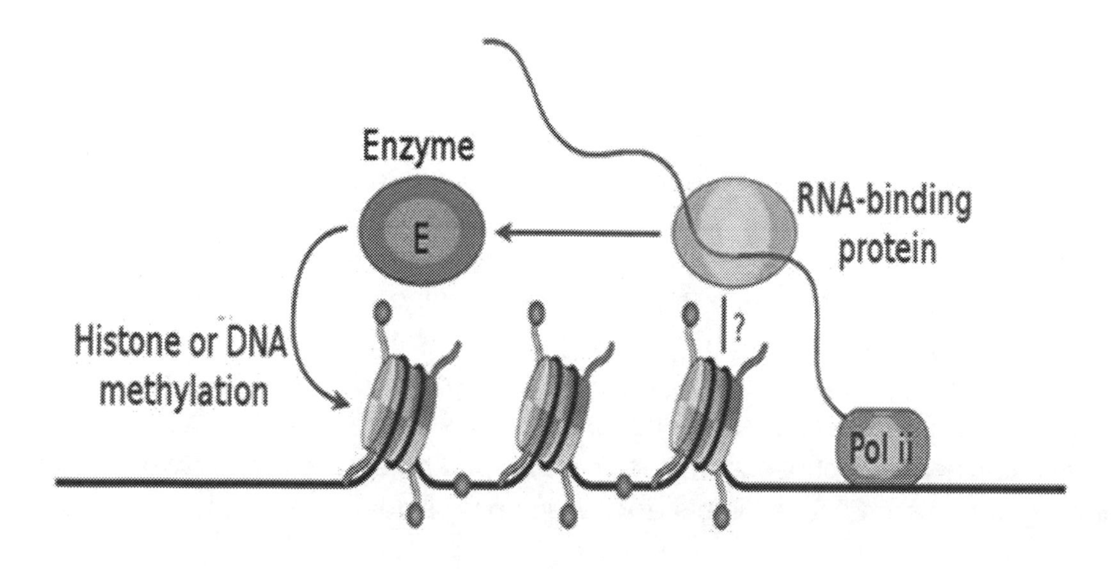


## 2.3.9 Phosphorylation of proteins

Phosphorylation is the addition of a phosphoryl group ($PO_3^{2-}$) to a molecule. Phosphorylation and its counterpart, dephosphorylation, turn many protein enzymes on and off, thereby changing their function and activity. Protein phosphorylation is one type of post-translational modification.

Protein phosphorylation in particular plays a significant role in a wide range of cellular processes. Its prominent role in biochemistry is the subject of a very large body of research (as of March 2015, the Medline database returns over 240,000 articles on the subject, largely on *protein* phosphorylation).

Phosphorylation refers to the addition of a phosphate to one of the amino acid side chains of a protein. Remember that proteins are composed of amino acids bound together and that each amino acid contains a particular side chain, which distinguishes it from other amino acids. Phosphates are negatively charged (with each phosphate group carrying two negative charges) so that their addition to a protein will change the characteristics of the protein. This change is often a conformational one, causing the protein to change how it is structured. This reaction is reversible by a process called dephosphorylation. The protein switches back to its original conformation when the phosphorus is removed. If these two conformations provide the protein with different activities (i.e. being enzymatically active in one conformation but not the other), phosphorylation of the protein will act as a molecular switch, turning the activity on or off. The transfer of phosphates onto proteins is catalyzed by a variety of enzymes in the cell. Although the variety is large, all of these enzymes share certain characteristics and fall into one class of proteins, called protein kinases. Their similarities stem from the group's ability to take a phosphate off the chemical energy-carrying molecule ATP and place it onto an amino acid side chain of a protein. The hydroxyl groups (-OH) of serine, threonine, tyrosine or histidine amino acid side chains are the most common target. A second class of enzymes is responsible for the reverse reaction, in which phosphates are removed from a protein. These are termed protein phosphatases, Sefton BM, Hunter T. 1998. Protein Phosphorylation. San Diego: Academic Press.

Within a protein, phosphorylation can occur on several amino acids. Phosphorylation on serine is the most common, followed by threonine. Tyrosine phosphorylation is somewhat rare but is at the origin of protein phosphorylation signaling pathways in most of the eukaryotes. However, since tyrosine phosphorylated proteins are relatively easy to purify using antibodies, tyrosine phosphorylation sites are relatively well understood. Histidine and aspartite phosphorylation occurs in prokaryotes as part of two-component signaling and in some cases in eukaryotes in some signal transduction pathways.

The cell cycle consists of a G1 growth phase, a S phase where the DNA of the cell is replicated (doubled), a G2 second growth phase, and ultimately a M (mitosis) phase where the cell divides forming a new cell. Certain regulatory molecules bind to cell membrane receptors that have an intercellular domain that is called a tyrosine kinase (tyrosine is one of the 20 amino acids that are used by cells to manufacture proteins and kinase means phosphotransferase). A tyrosine kinase is an enzyme that can transfer a phosphate group from ATP to a tyrosine residue in a protein as the following equation:

ATP + protein tyrosine = ADP + protein tyrosine phosphate

Many enzymes and receptors are switched "on" or "off" by phosphorylation and dephosphorylation. Scientists worked out most of the details of how ras proteins control the operation of the cell cycle. It was subsequently found that single point mutation of the ras gene can lead to the formation of an oncogene and a cancer causing ras protein.

Ras proteins are monomeric G-proteins of widespread importance.... a protein kinase cascade, which ultimately leads to the phosphorylation of transcription factors in the nucleus, which in turn alter gene expression. The binding of a growth factor molecule to a receptor on the cell membrane is the stimulus that begins the process. The ras protein is normally in the inactive state when it is associated with a molecule called GDP. Upon being activated by the growth factor stimulus, it switches GDP for the more active GTP (GDP and GTP stand for guanosine diphosphate and guanosine triphosphate nucleotide). Before the ras protein can become functional, it must receive a small molecule with the aid of an enzyme called farnesyl transferase (Farnesyl transferase is responsible for activating RAS). At this point, the ras protein attaches to the cell membrane and initiates a cascade of enzymatic reactions. Eventually a protein enters the nucleus, where it activates a “transcription factor”, which in turn activates the cyclin D gene. The cyclin D protein that is formed stimulates the progression of the cell cycle into the S phase and through the remainder of the cycle. This cyclin forms a complex with and functions as a regulatory subunit of cyclins group whose activity is required for cell cycle G1/S transition. In normal cells, the ras protein reverts to the inactive state after transmitting the signal by replacing GTP to GDP, Figure 38. However, a mutation in the ras gene results in the formation of a ras protein that no longer has the enzyme to convert GTP to GDP. Therefore, the ras protein remains in the active state even with no stimulus from growth factors. The ras protein then sends continuous signals to keep the cell cycle running with no checks and balances. The result is excessive cell proliferation and cancer, Figure (42).

Figure (42): Ras protein alters gen expression

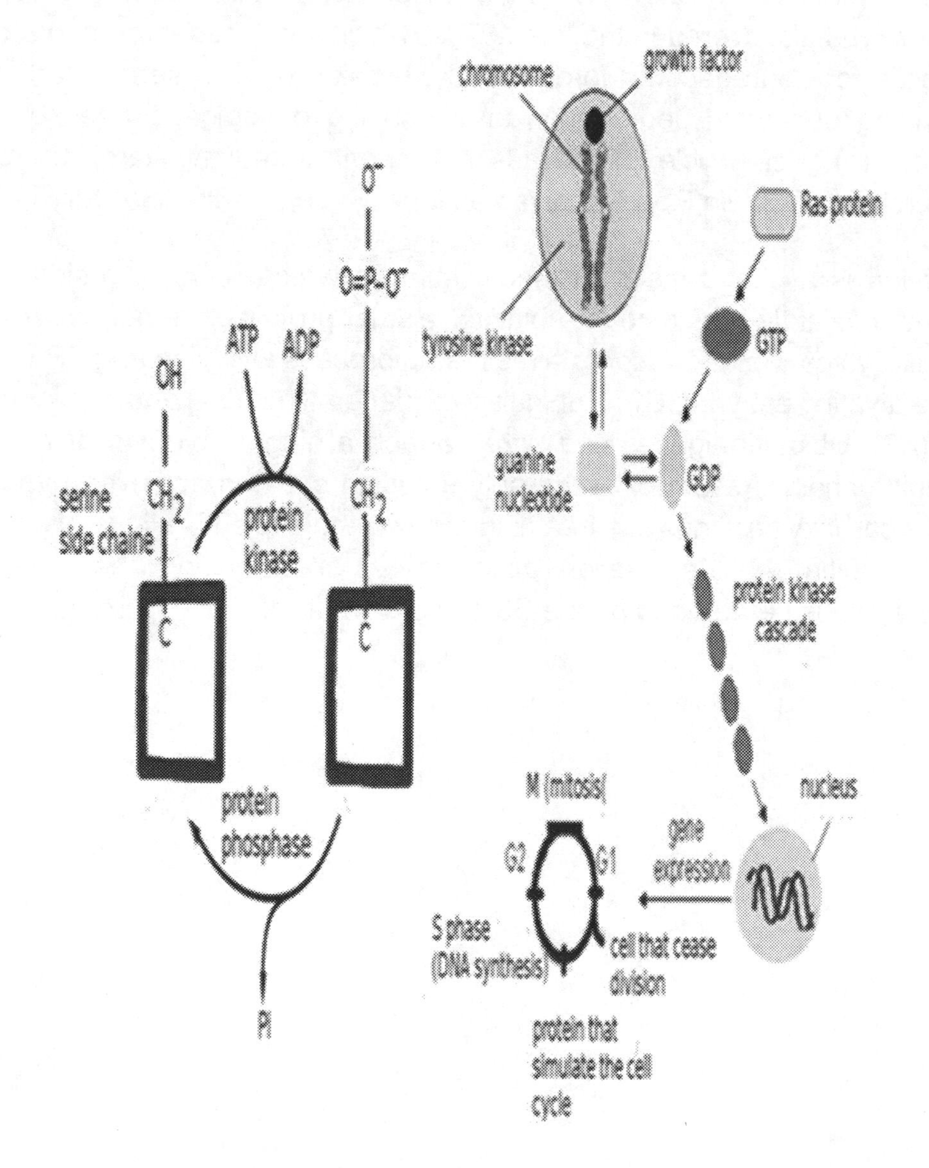

## 2.3.10 Ubiquitylation of Protein

Ubiquitin is a small (8.5 kDA) regulatory protein that has been found in almost all tissues of eukaryotic organisms. It was found in 1975 by Gideon Goldstein and further characterized throughout the 1970s and 1980s. There are four genes in the human genome that produce ubiquitin: UBB, UBC, UBA52 and RPS27A.

Ubiquitylation, also known as Ubiquitination, is an enzymatic process that involves the bonding of an ubiquitin protein to a substrate protein. Ubiquitin and at least ten ubiquitin-like proteins are important post-translational modifiers that regulate nearly every aspect of cellular function. This has sometimes been referred to as the molecular "kiss of death" for a protein, as the substrate usually becomes inactivated and is tagged for degradation by the proteasome (a protein complex in cells containing proteases; it breaks down proteins that have been tagged by ubiquitin) through the attachment of the ubiquitin molecule. The theory of ubiquitylation involving the labeling of proteins for degradation was discovered in the late 1970s, which led to research about targeting specific cells for degradation. More recently, breakthrough research furthered the understanding of protein degradation and resulting biological processes, making possible control of cell cycle, gene transcription and immunity. Aaron Ciechanover, Avram Hershko and Irwin Rose received a Nobel Prize for this discovery in 2004.

Ubiquitylation is a post-translation modification in which a lysine residue (K) of a substrate is covalently conjugated to ubiquitin, a small protein extremely well conserved among eukaryotes. Ubiquitylation is a multi-step process involving an ATP-dependent ubiquitin-activating enzyme (E1), a ubiquitin-carrier enzyme (E2) and a ubiquitin-ligase enzyme (E3). Ubiquitin-ligases can either attach a single ubiquitin or assemble a poly-ubiquitin chain. Assembly of the poly-ubiquitin chain is achieved by covalently binding the carboxy terminus of a free ubiquitin molecule to a K residue of a previously attached ubiquitin (there are seven 'acceptor' K's on ubiquitin). The substrate and chain specificity is determined by the E3 alone or with the E2, Figure (43).

Figure (43): Enzymes E1. E2, and E3 and ubiquitylation of protein

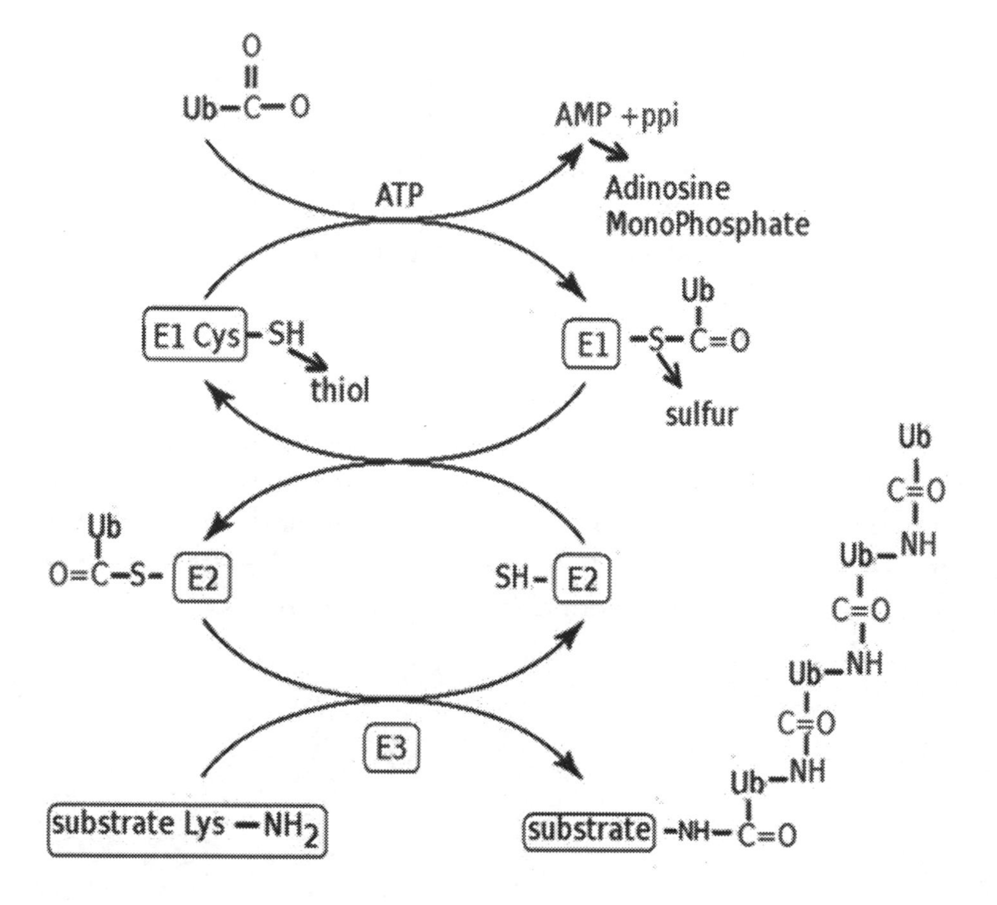


Different types of ubiquitylation establish alternative fates for substrates targeted by E3 ligases.

The substrates targeted by Cbl proteins (Cbl is an E3 ubiquitin-protein ligase involved in cell signalling and protein ubiquitination) can either be multi-ubiquitylated, or poly-ubiquitylated, as is the case for Sprouty proteins. The mechanisms that enable Cbl proteins to direct both types of ubiquitylation remain to be concluded. Cbl is a mammalian gene encoding the protein CBl which is an E3 ubiquitin-protein ligase involved in cell signaling and protein ubiqitination. Mutations to this gene have been implicated in a number of human cancers, particularly acute myeloid leukemia.

### 2.3.11 Sumoylation

Small Ubiquitin-like Modifier (or SUMO) proteins are a family of small proteins that are covalently attached to and detached from other proteins in cells to modify their function. Sumoylation is a post-translational modification process. It is analogous to ubiquitylation in terms of the reaction scheme and enzyme classes used, but rather than conjugation by ubiquitin, sumoylation involves addition of SUMOs (small ubiquitin-like modifiers).

Small Ubiquitin-like Modifier (or SUMO) proteins are a family of small proteins that are covalently attached to and detached from other proteins in cells to modify their function. SUMOylation is a post-transational modification involved in various cellular processes, such as nuclear-cytosolic- transport, transcriptional regulation, apoptosis, protein stability, response to stress, and progression through the cell cycle.

### 2.3.12 ADP Ribosylation

ADP-ribosylation is the addition of one or more ADP-ribose moieties to a protein Adenosine diphosphate ribose is an ester (ester is a molecule with one double bonded oxygen and one single bonded oxygen) molecule formed into chains by the enzyme poly ADP ribose polymerase. It binds to and activates the TRPM2 (a protein that in humans is encoded by the TRPM2 gene) ion channel, Figure (44).

Figure (44): ADP ribosylation

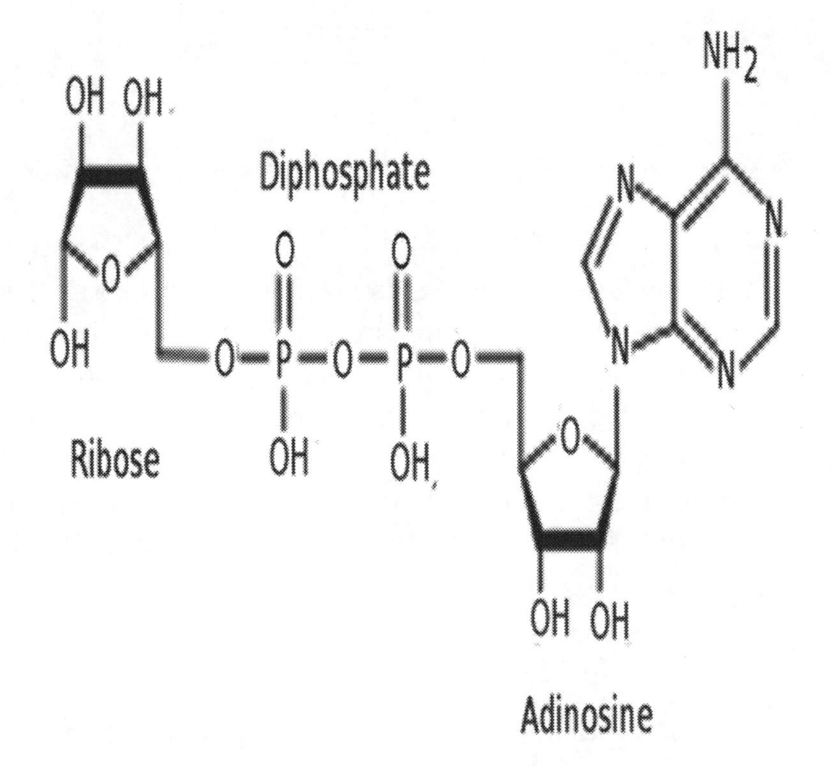

it became clear that ADP-ribosylation reactions play important roles in a large range of physiological and pathophysiological processes, including inter- and intracellular signaling, transcriptional regulation, DNA repair pathways and maintenance of genomic stability, telomere dynamics, cell differentiation and proliferation, and necrosis and apoptosis. ADP-ribosylation is a posttranslational modification of proteins that engages the addition of one or more ADP and ribose moieties. These reactions are involved in cell signaling and the control of many cell processes.

ADP-ribosylation is also responsible for the actions of some bacterial toxins, such as cholera toxin, diphtheria toxin, and pertussis toxin. These toxin proteins are ADP-ribosyltransferases that modify target proteins in human cells. For example, cholera toxin ADP-ribosylates G proteins, which causes massive fluid secretion from the lining of the small intestine and results in life-threatening diarrhea.

This protein posttranslational modification is produced by ADP-ribosyltransferase enzymes, which transfer the ADP-ribose group from nicotinamide adenine dinucleotide (NAD+) onto acceptors such as arginine, glutamic acid or aspartic acid residues in their substrate protein. In humans, one type of ADP-ribosyltransferases are the NAD: arginine ADP-ribosyltransferases, which modify amino acid residues in proteins such as histones by adding a single ADP-ribose group.

These reactions are reversible; for example, when arginine is modified, the ADP-ribosylarginine produced can be removed by ADP-ribosylarginine hydrolases.

As well as the transfer of single ADP-ribose moieties, multiple groups can also be transferred to proteins to form long branched chains, in a reaction called poly(ADP-ribosylation. This protein modification is carried out by the poly ADP-ribose polymerases (PARPs), which are found in most eukaryotes, but not prokaryotes or yeast. The poly(ADP-ribose) structure is involved in the regulation of several cellular events and is most important in the cell nucleus, in processes such as DNA repair and telomere maintenance, Figure (45).

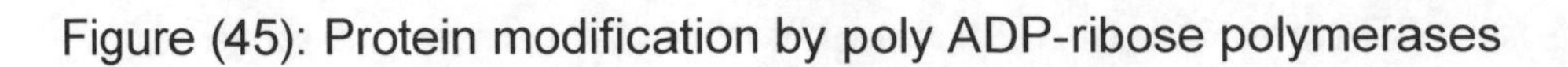
Figure (45): Protein modification by poly ADP-ribose polymerases

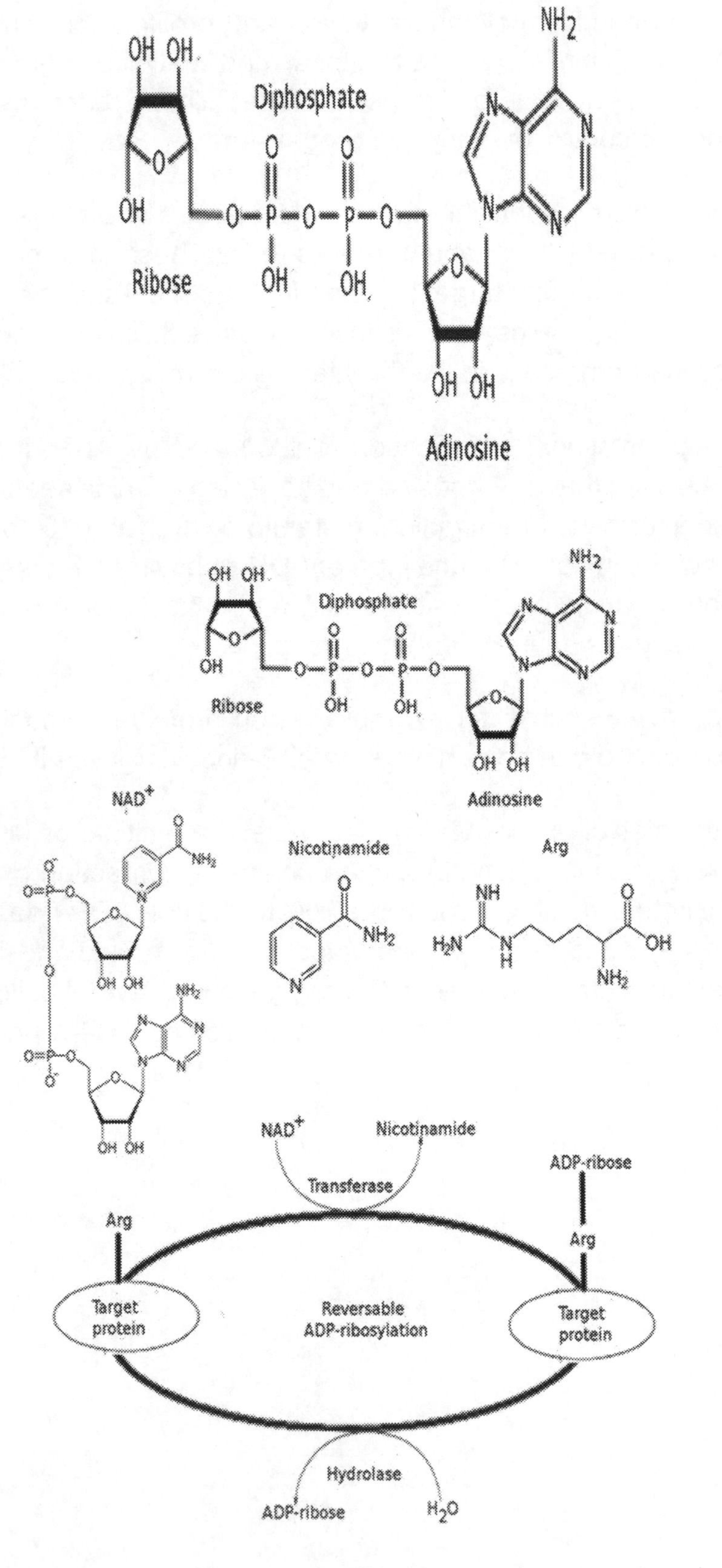

### 2.3.13 Deimination

Citrullination or deimination is the conversion of arginine to citrulline, identified as citrullination or deimination.. Enzymes called peptidylarginine deiminases (PADs) replace the primary ketimine group (=NH) by a ketone group (=O), Figure (46).

Figure (46): Deimination

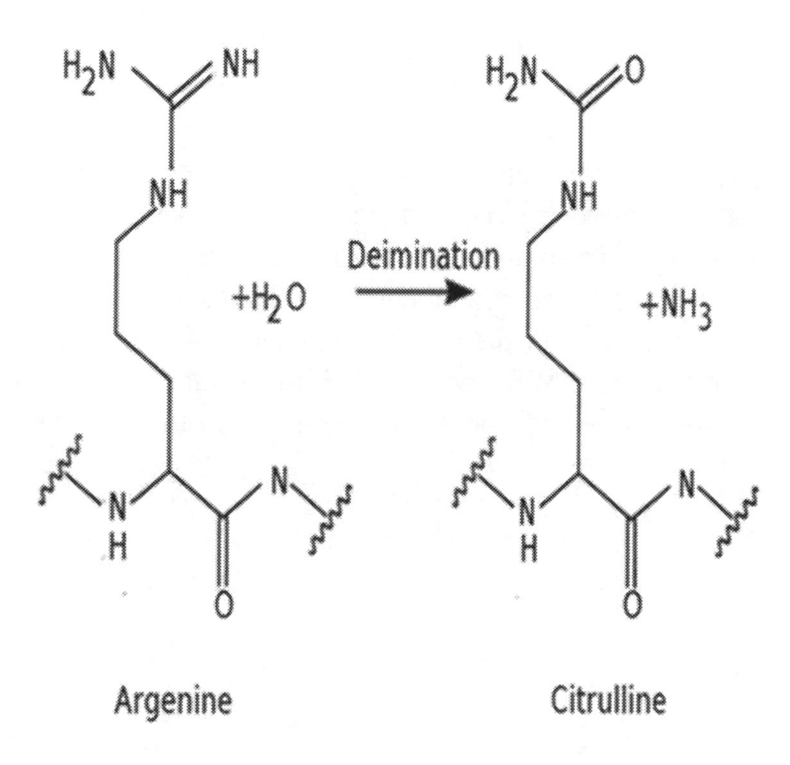


The elimination of hydrogen atom in the molecule of argentine will build up increase the positive charge and consequently this increases the hydrophobicity of the protein, leading to changes in protein folding. Therefore, citrullination can change the structure and function of proteins.

Citrullination controls the expression of genes, particularly in the developing embryo. The immune system often attacks citrullinated proteins, leading to autoimmune diseases such as rheumatoid arthritis and multiple sclerosis.

Citrulline is not one of the 20 standard amino acids encoded by DNA in the genetic code. Instead, it is the result of a post-translational modification.

Peptidylarginine Deiminase 4 (PAD4) is an enzyme that converts Arg or monomethyl-Arg to citrulline in histones, is essential for decondensed chromatin. The areas of broad chromatin decondensation along the deimination were rich in histone citrullination. Here, upon investigating the effect of global citrullination in cultured cells, it was found that PAD4 overexpression in osteosarcoma U2OS cells induces extensive chromatin decondensation independent of apoptosis. The highly decondensed chromatin is released to the extracellular space and stained strongly by a histone citrulline-specific antibody. The formation of the decondensed chromatin is reminiscent of deimination but is unique in that it occurs without stimulation of cells with pro-inflammatory cytokines and bacteria. Furthermore, histone citrullination during chromatin decondensation can dissociate heterochromatin protein 1 beta (HP1β) thereby offering a new molecular mechanism for understanding how citrullination regulates chromatin function. Taken together, it suggests that PAD4 mediated citrullination induces chromatin decondensation, implicating its important role in NET formation under physiological conditions in neutrophils. Neutrophil extracellular traps (NETs) are an innate immune defense mechanism, yet NETs also may aggravate chronic inflammatory and autoimmune disorders. Activation of peptidylarginine deiminase 4 (PAD4) is associated with NET release (NETosis) but the precise mechanisms of PAD4 regulation are unfamiliar. The PAD proteins are only activated when sufficient Ca2+ is available, Figure (47).

Figure (47): Effect of PAD4 on protein condensation

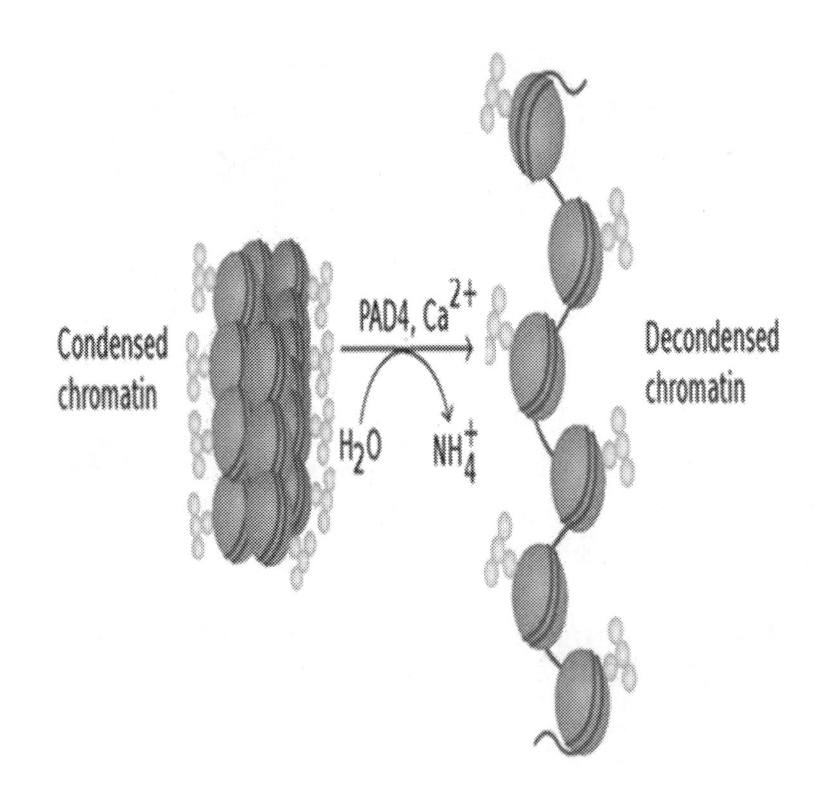

### 2.3.14 Biotinylation

In biochemistry, biotinylation is the process of covalently attaching biotin to a protein, nucleic acid or other molecule. Biotinylation is rapid, specific and is unlikely to perturb the natural function of the molecule due to the small size of biotin (MW = 244.31 g/mol). The binding between biotin (Biotin, also known as vitamin H or vitamin B7) and streptavidin, avidin or Neutravidin is one of the strongest known non-covalent biological interactions. The (strept)avidin-biotin interaction has been commonly used for decades in biological research and biotechnology. Therefore labeling of purified proteins by biotin is a powerful way to achieve protein capture, immobilization, and functionalization, as well as multimerizing or bridging molecules. Biotin-binding to streptavidin and avidin is resistant to extremes of heat, pH and proteolysis, making capture of biotinylated molecules possible in a wide variety of environments.

One example of amplification of an antibody is the immunohistochemical (IHC) staining intensity which is a function of the enzyme activity, and improved sensitivity can be done by increasing the number of enzyme molecules bound to the tissue. The multiple binding sites between the tetravalent avidin and biotinylated antibodies (bound to the antigen) are ideal for achieving this amplification. The two most common for amplifying the target antigen signal in IHC are called avidin-biotin complex (ABC) and labeled streptavidin binding, Figure (48).

The multiple biotin binding sites in each tetravalent avidin molecule are ideal for achieving this amplification. The information below describes the general procedure.

1. The primary antibody is incubated with the tissue sample to le binding to the target antigen.
2. A biotinylated secondary antibody, with specificity against the primary antibody, is incubated with the tissue sample to allow binding to the primary antibody.
3. A biotinylated enzyme (HRP or AP) is pre-incubated with free avidin to form large avidin-biotin-enzyme complexes. Typically, the avidin and biotinylated enzyme are mixed together in a specified ratio to prevent avidin saturation..
4. An aliquot of this solution is then added to the tissue sample, and any remaining biotin-binding sites on the avidin bind to the biotinylated antibody that is already bound to the tissue.

Figure (48): Biotinylation

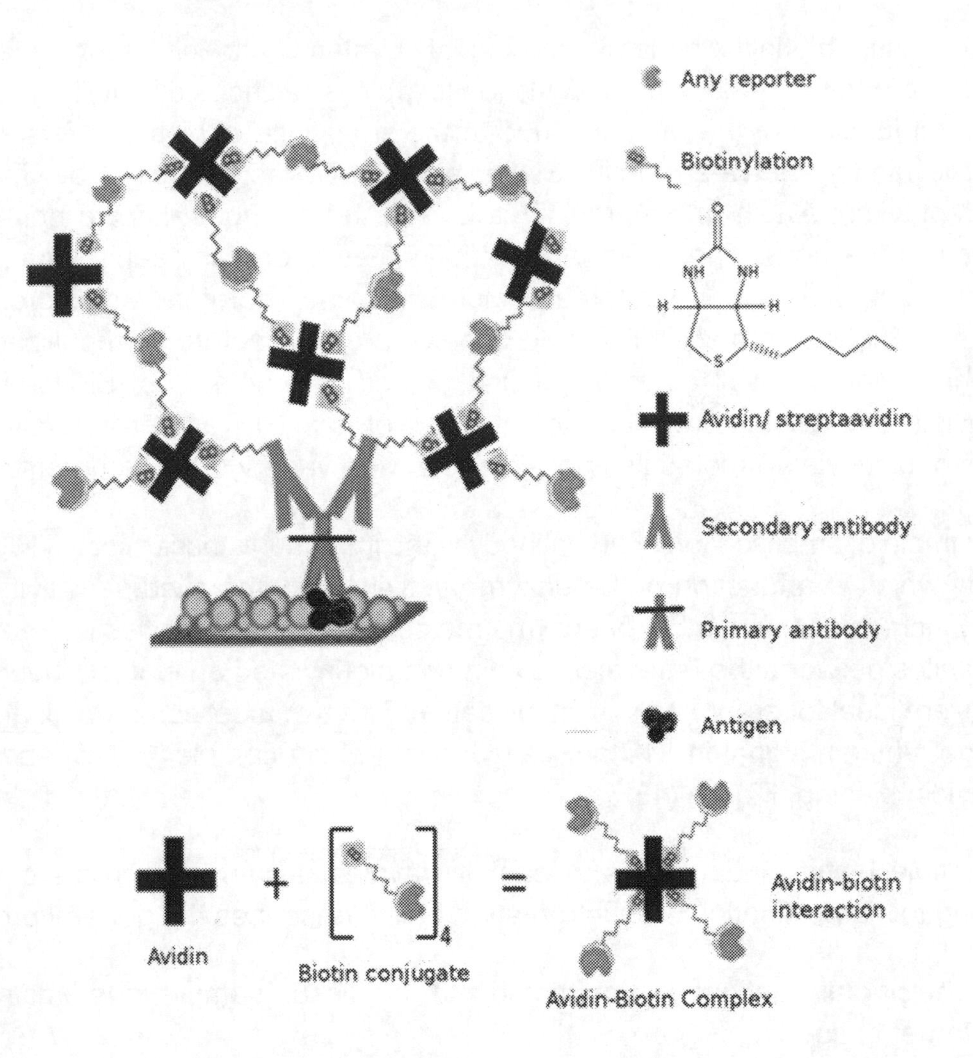


## 2.3.15 Propionylation and Butyrylation

Lysine can be modified in histones, lysine propionylation and butyrylation. Recent studies of protein post-translational modifications showed that various types of lysine acylation occur in eukaryotic and bacterial proteins. Lysine propionylation and Butyrylation, newly discovered types of acylation, occurs in several proteins, including some histones. Several proteins interact either to activate or repress the expression of other genes during transcription. Based on the impact of these activities, the proteins can be categorized classified into readers, modifier writers, and modifier erasers depending on whether histone marks are read, added, or removed, respectively, from a specific amino acid. Transcription is controlled by dynamic epigenetic marks with serious health implications in certain complex diseases, whose understanding may be useful in gene therapy, Figure (49).

Figure (49): Propionylation and Butyrylation

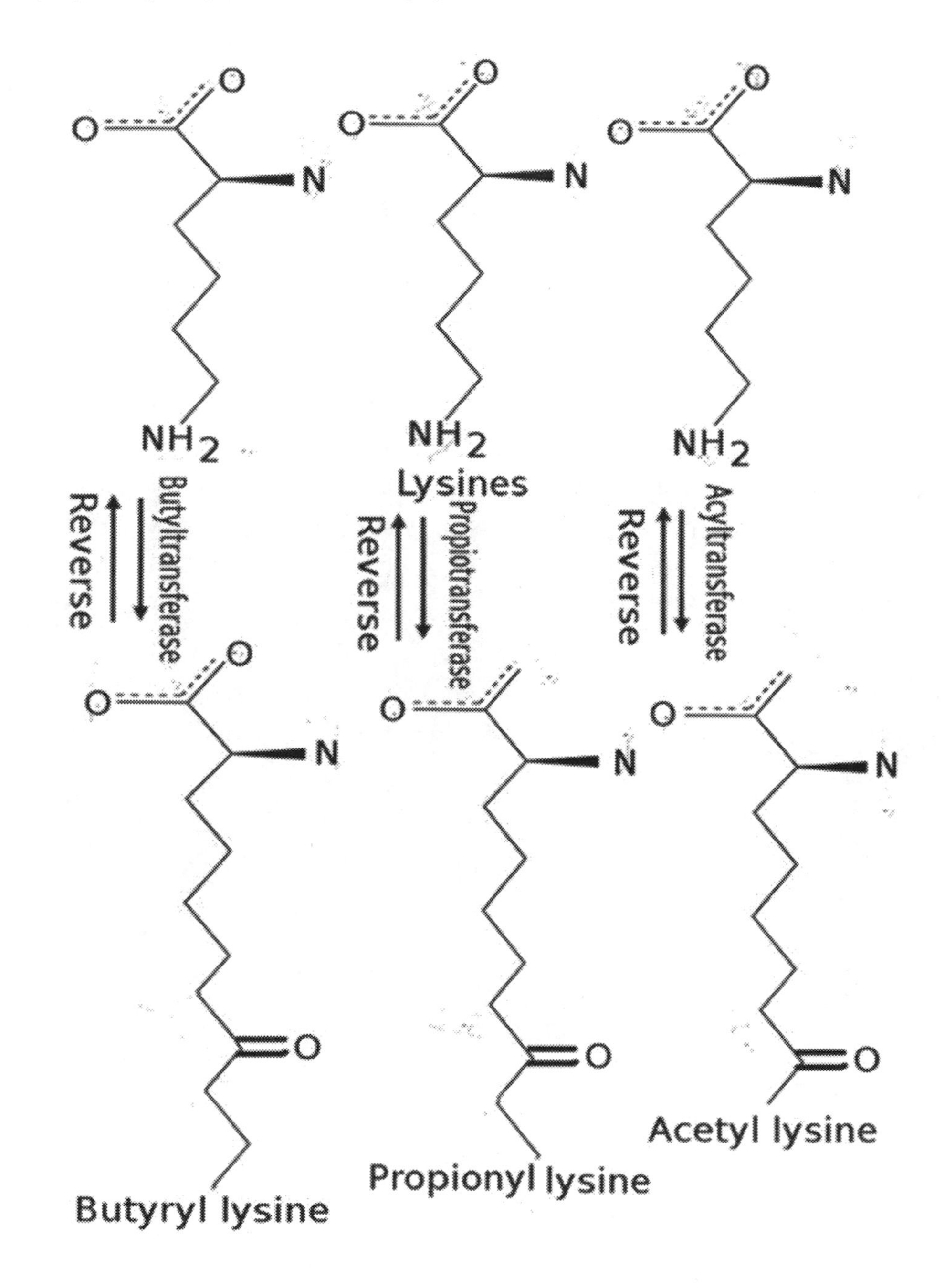


Lysine is encoded by the codons AAA and AAG). It is a α-amino acid that is used in the biosynthesis of proteins. Lysine is a base, as are argenine and histidine. The ε-amino group often participates in hydrogen bonding and as a general base in catalysis.

## 2.3.16 N-formylation

In chemistry, the addition of a formyl functional group is termed formylation. A formyl functional group consists of a carbonyl bonded to a hydrogen. When attached to an R group, a formyl group is called an aldehyde.

N-Formyl compounds have been widely used in organic synthesis as protecting group of amines. Formamide synthesis is one of the important areas in organic chemistry as they contribute to be the key intermediate in synthesis of various pharmaceutical compounds. Formamides also act as an amino-protecting group during peptide synthesis and as precursors in preparations of *N*-methyl compounds. Various methods have been developed for formylation of amines in recent years, Figure (50). However, many of these methods suffer from drawbacks such as use of expensive and toxic formylating agents or employing expensive catalyst, formation of side products, use of excessive amount of formylating reagent/substrate ratio, etc.

Figure (50): Formylation of amines

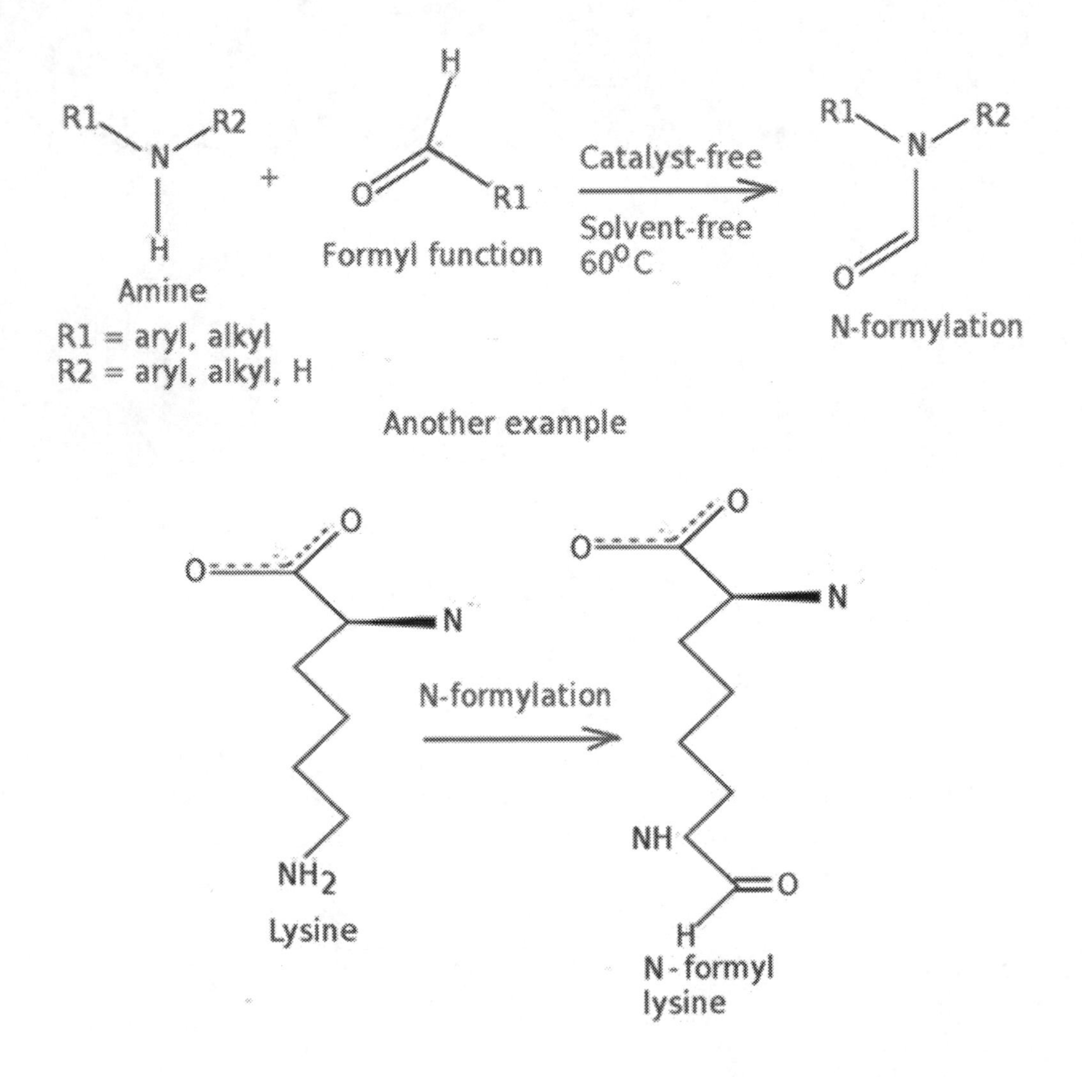

In histone proteins, lysine is typically modified by Histone Acetyl-Transferases (HATs) and Histone Deacetylases (HDAC or KDAC). The acetylation of lysine is fundamental to the regulation and expression of certain genes. Oxidative stress creates a significantly different environment in which acetyl-lysine can be quickly outcompeted by the formation of formyl-lysine due to the high reactivity of formylphosphate species. This situation is currently believed to be caused by oxidative DNA damage. A mechanism for the formation of formylphosphate has been proposed, which it is highly dependent on oxidatively damaged DNA and mainly driven by radical chemistry within the cell[1]. The formylphosphate produced can then be used to formylate lysine. Oxidative stress is believed to play a role in the availability of lysine residues in the surface of proteins and the possibility of being formylated, Figure (51).

1- Jiang, T; Zhou, X.; Taghizadeh, K.; Dong, M.;, Dedon, PC. (2007). N-formylation' N-formylation of lysine 'N-formylation PNAS 104 (1): 60–*65*.doi:10.1073/pnas.0606775103. PMC 1765477 PMID 17190813. Retrieved 24 February 2013.

Figure (51): Formulation of lysine of proteins

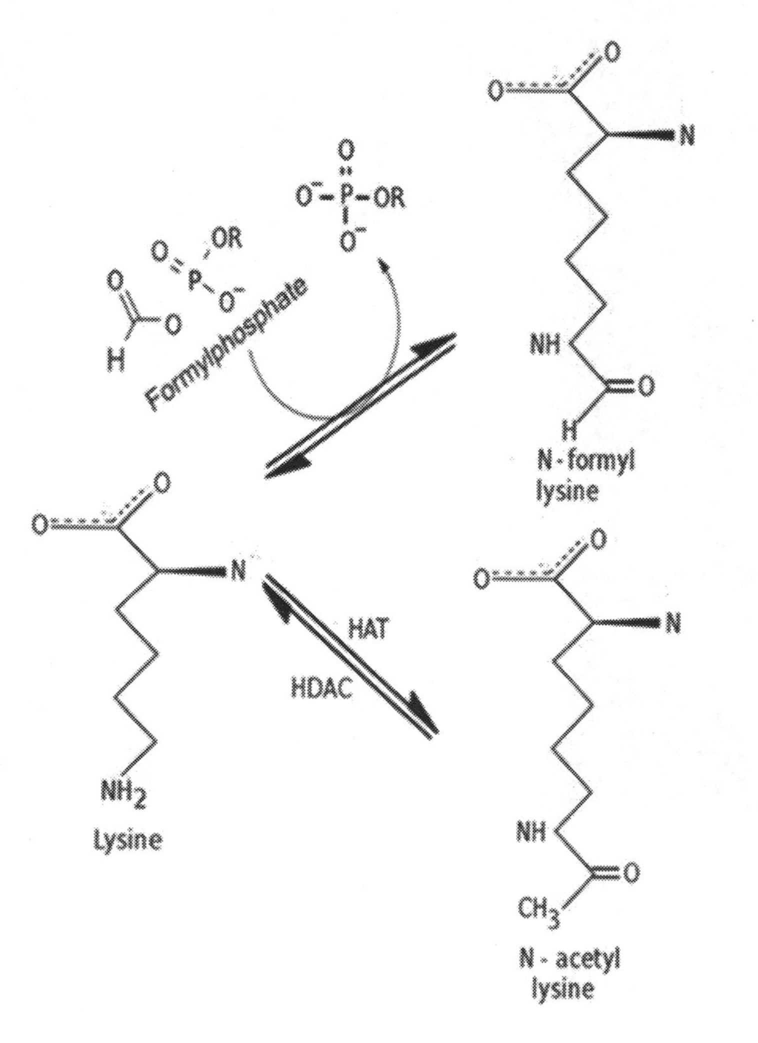

## 2.3.17 Proline Isomerization

Proline is encoded by the codons CCU, CCC, CCA, and CCG. It is non-essential in humans, meaning the body can synthesize it from the non-essential amino acid L-glutamate. It is an alpha-amino acid that is used in the biosynthesis of proteins. In biochemistry, amino acids having both the amine and the carboxylic acid groups attached to the first (alpha) carbon atom have particular importance, beta with the carboxylic acid attached to the second carbon and Gamma to the third carbon.

Peptide bonds to proline, and to other *N*-substituted amino acids are able to populate both the cis and trans isomers. Most peptide bonds overwhelmingly adopt the trans isomer (93%) and about 7% fir cis, Figure (52).

Figure (52): Proline Isomeration

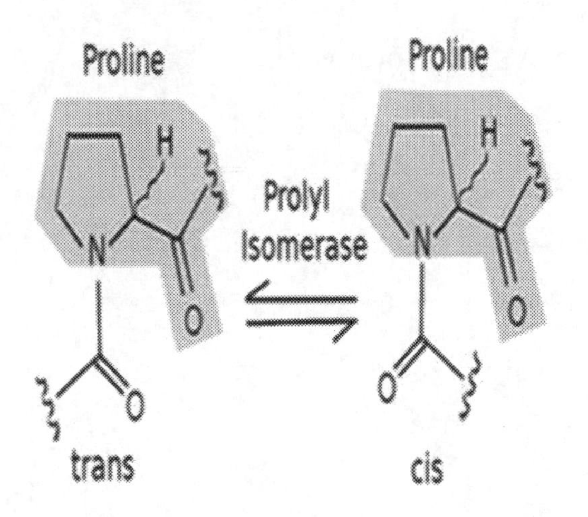


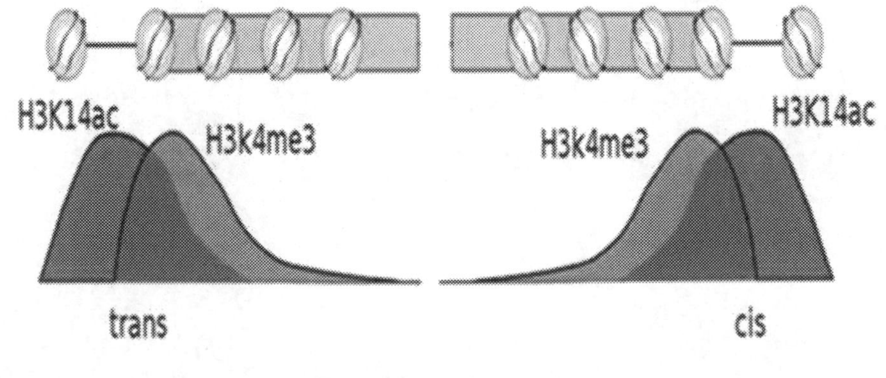

The cis-trans isomerization of proline serves as a regulatory switch in signaling pathways. Proline isomerization can be used as a novel noncovalent histone modification that regulates transcription and provides evidence for crosstalk between histone lysine methylation and proline isomerization.

CHAPTER 3

# GENE EDITING

Gene editing, Genome editing, or genome editing with engineered nucleases (GEEN) is a type of genetic engineering in which DNA is inserted, deleted or replaced in the genome of an organism using engineered nucleases, or "molecular scissors." These nucleases create site-specific double-strand breaks (DSBs) at desired locations in the genome. A nuclease is an enzyme capable of cleaving the phosphodiester bonds between the nucleotide subunits of nucleic acids.

There are currently four families of engineered nucleases being used: Meganucleases, Zinc finger nucleases (ZFNs), Transcription Activator-like Effector-based Nucleases (TALEN), and the CRISPR-Cas-Cas system.

Genome editing was selected by Nature Methods as the 2011 Method of the Year. The CRISPR-Cas system was selected by Science Magazine as 2015 Breakthrough of the Year.

The ethical and societal issues surrounding heritable gene editing recently drew more than 400 scientists, bioethicists and historians of science from 20 countries to Washington, D.C., for the International Summit on Human Gene Editing, hosted by the National Academy of Sciences, the National Academy of Medicine, the Chinese Academy of Sciences, and The Royal Society of the United Kingdom.

These are the potential dangers of making changes to the human genome that can be passed down to future generations, and an issue that has become more urgent with the advent of CRISPR-Ca9, an easy-to-use and cheap way to precisely edit animal and plant genomes.

Gene editing will give scientists the ability to quickly and simultaneously make multiple genetic changes to a cell. Many human illnesses, including heart disease, diabetes, and assorted neurological conditions, are affected by numerous variants in both disease genes and normal genes.

Mainly, there are two types of gene editing:

- Cell"s own repair mechanisms in which the cell performs the repair. Once a break is introduced in the DNA, the cell will detect a problem in its genetic code and quickly activate its repair machinery. When the DNA (or RNA) breaks, the cell can employ different enzymes to directly join the two ends of the DNA break back together. This process, known as nonhomologous end-joining, is very error-prone and often results in mutations – such as small insertions or deletions of nucleotides – in the resulting DNA strand. These small mutations can be neutral, but they can also render the entire gene in that location nonfunctional, achieving the disruption or knockout of the gene.
- In vitro/in vivo manual editing in which scientists can directly change genetic information in cells by presenting a correct version of a DNA sequence to replace an unwanted mutation.

Both approaches could show useful in the context of the treatment of many diseases. However, there must be a way to direct a nuclease to the desired location where a DNA break is to be introduced. To address this issue, many different types of nucleases have been utilized. All nucleases consist of 2 components – the nuclease itself that is responsible for DNA cleavage and a secondary component responsible for recognizing a specific DNA sequence. There are three main classes of nucleases engineered for genomic editing purposes:

## 3.1 Transcription activator-like effector nucleases (TALENs)

TALENs is a widely applicable technology for targeted genome editing.

TAL (transcription activator-like) effectors (often referred to as TALEs) are proteins secreted by Xanthomonas bacteria via their type III secretion system when they infect different plant species. These proteins can bind promoter sequences in the host plant' DNA and activate the expression of plant genes that aid bacterial infection. The DNA-binding domain of each TALE consists of tandem 34–amino acid repeat modules that can be rearranged according to a simple cipher to target new DNA sequences. Usually, two hypervariable amino acid residues in each repeat recognize one base pair in the target DNA. There seems to be a one-to-one correspondence between the identity of two critical amino acids in each repeat and each DNA base in the target sequence. These proteins are interesting to researchers both for their role in disease of important crop species and the relative ease of retargeting them to bind new DNA sequences. Similar proteins can be found in the pathogenic bacterium Ralstonia solanaserum and Bukholderia rhizoxinica. TALE nuclease can cleave double-stranded DNA in vitro if the DNA binding sites have the proper spacing and orientation.

Transcription activator-like effectors (TALEs) are important Xanthomonas virulence factors that bind DNA via a unique tandem 34-amino-acid repeat domain to induce expression of plant genes. So far, TALE repeats are described to bind as a consecutive array to a consecutive DNA sequence, in which each repeat independently identifies a single DNA base. This modular protein architecture makes possible the design of any desired DNA-binding specificity for biotechnology applications. Natural TALE repeats of unusual amino-acid sequence length break the strict one repeat-to-one base pair binding mode and introduce a local flexibility to TALE–DNA binding. This flexibility allows TALEs and TALE nucleases to recognize target sequence variants with single nucleotide deletions. The flexibility also allows TALEs to activate transcription at allelic promoters that otherwise give resistance to the host DNA, Figure (53).

Figure (53): TALE–DNA binding

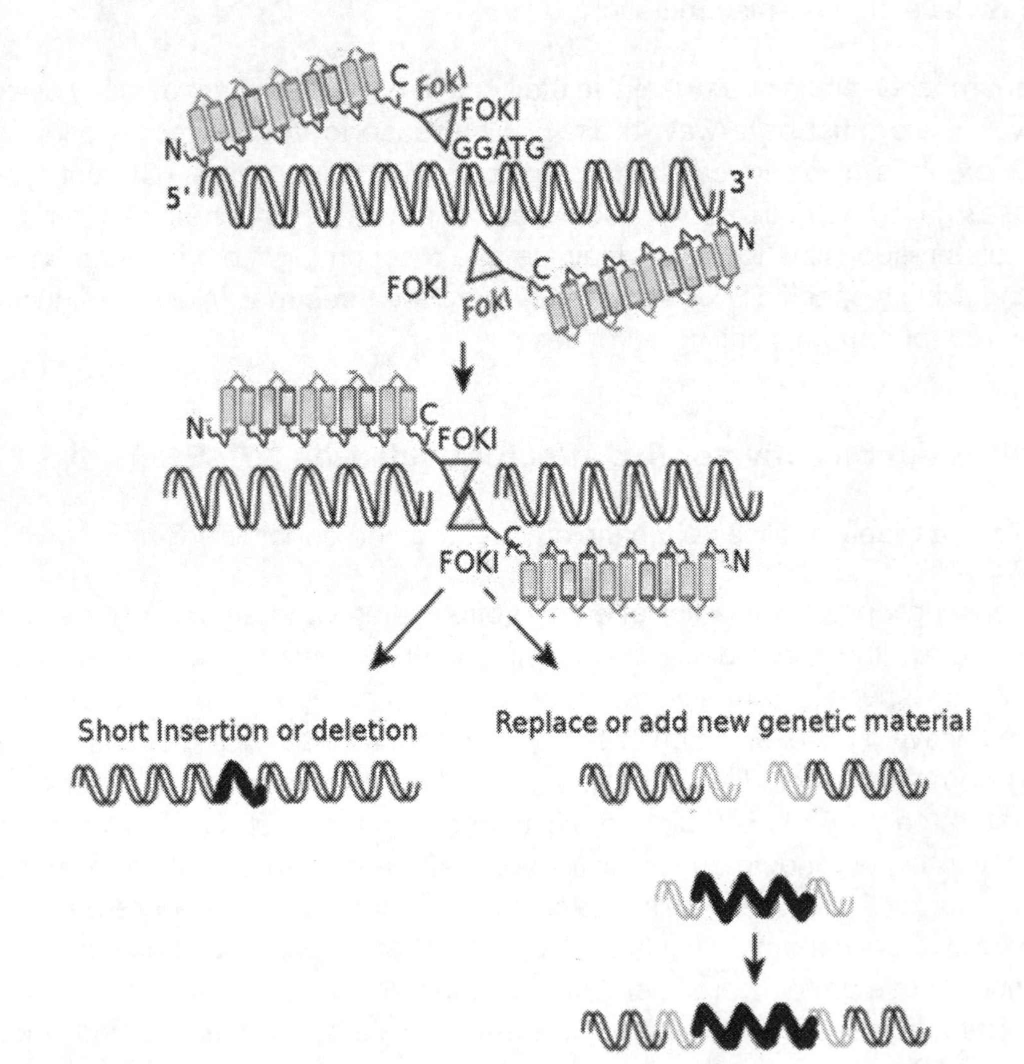

The enzyme FokI, naturally found in Flavobacterium okeanokoites, is a bacterial type IIS restriction endonucleaclease consisting of an N-terminal DNA-binding domain and a non-specific DNA cleavage domain at the C-terminal. Once the protein is bound to dublex DNA via its DNA-binding domain at the 5'-GGATG-3' recognition site, the DNA cleavage domain is activated and cleaves.

Zhang et al (2014)) explain their efforts to understand some of the rules for using TALEs as transcriptional activators or repressors. TALEs, because of their programmable DNA binding properties, can be linked to other functional protein domains besides nucleases to create DNA binding proteins with other functionalities. For example, if TALEs are fused with VP16/64 and programmed to bind upstream of a gene it can recruit transcription initiation complexes to initiate transcription, Figure (54).

Figure (54): TALE and possible functional domains

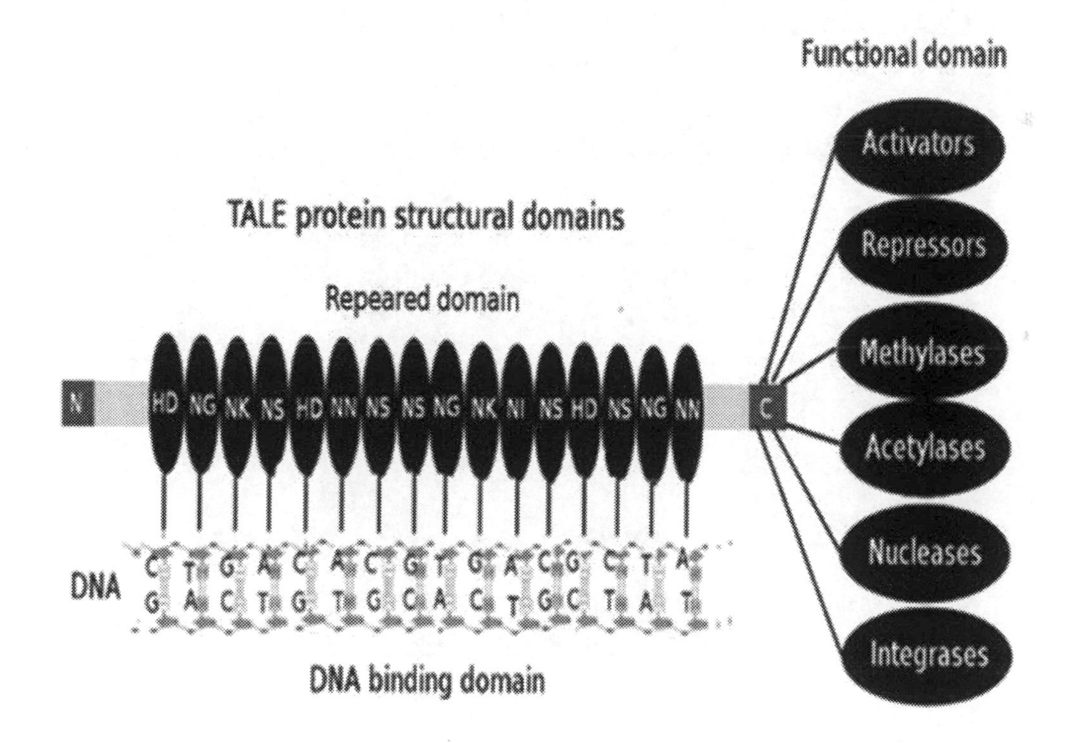


### 3.1.1 TALE Structure

TALE structure represents Hax3 and dHax3.N amino acids of TALEs and their DNA target specificity is shown in Figure (47). (A) Hax3 TALE has a central repeat domain as shown in Figure (55), a nuclear localization signal (NLS) is also, and a transcriptional activation domain (AD) shown in the same figure. The central repeat domain consists of 11.5 repeat units, and each repeat unit is composed of 34 amino acids; a representative first repeat sequence is shown below, and the RVD is also shown. (B) RVDs of the 11.5 repeat units of Hax3 and dHax3 with their DNA target sequences shown below. (C) Structural representation of the dHax3.N showing the fusion of 196 aa of the FokI cleavage domain at the C terminus of dHax3, Figure (55).

Figure (55): Domains of FokI cleavage

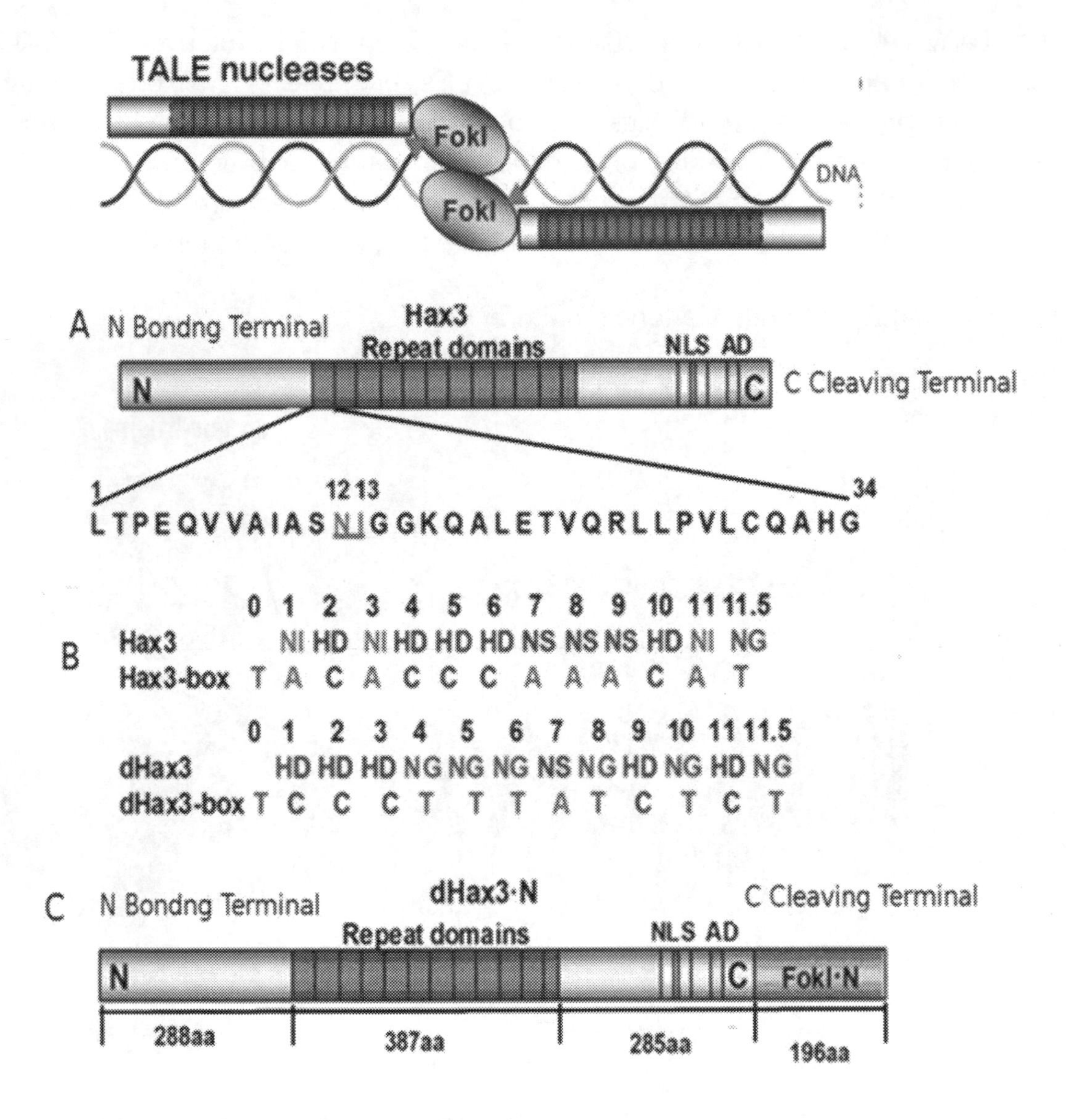


## 3.2 Zinc Finger Nucleases (ZFNs)

This application can target genome cleavage by engineered, sequence-specific zinc finger nucleases followed by gene modification during subsequent repair. Zinc finger domains can be engineered to target specific desired DNA sequences and this enables zinc-finger nucleases to target unique sequences within complex genomes.

Such 'genome editing' is now established in human cells and a number of model organisms, thus opening the door to a range of new experimental and therapeutic possibilities. Zinc-finger nucleases (ZFNs) are artificial restriction enzymes generated by fusing a zinc finger DNA-binding domain to DNA-cleavage domain. Zinc finger domains can be engineered to target specific desired DNA sequences and this enables zinc-finger nucleases to target unique sequences within complex genoms. By taking advantage of endogenous DNA repair machinery; these reagents can be used to precisely alter the genomes of higher organisms. ZFN is becoming a prominent tool in the field of genome editing.

A zinc finger nuclease is a site-specific endonuclease designed to bind and cleave DNA at specific positions. There are two protein domains. The first domain is the DNA binding domain, which consists of eukaryotic transcription and contains the zinc finger. The second domain is the nuclease domain, which consists of the FokI restriction enzyme and is responsible for the catalytic cleavage of DNA.

Individual zinc fingers are shown as small ovals, numbered 1, 2 and 3 starting from the N terminus, each contacting three base pairs, Figure (56).

Figure (56): Zinc fingers

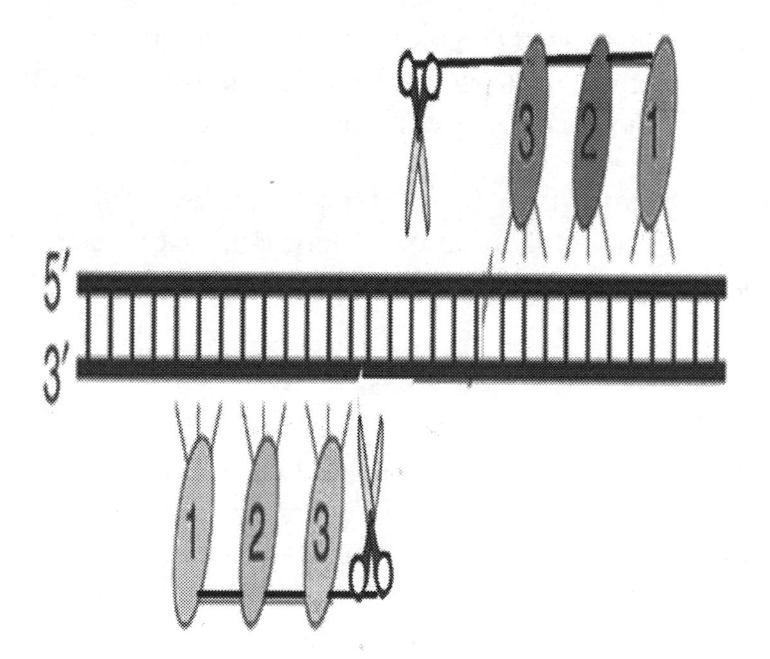


Majority of TFs consist of one or more DNA binding domains (DBDs) that provide sequence specificity or selectivity to the TFs allowing them to recognize their genomic targets. DNA binding domains have been categorized into different families based on their sequence homology. DBDs within a family also share basic mechanism of DNA recognition. Although TFs in the human genome can be divided into more than 20 groups based on their DBD, over 80% fall into three groups of DBDs namely, C2H2-Zinc finger domain (ZFD), Homeodomains (HD) and basic helix-loop-Helix domain (bHLH)*.

C2H2-type (classical) zinc fingers (Znf) were the first class to be characterised. They contain a short beta hairpin and an alpha helix (beta/beta/alpha structure) where a single zinc atom is held in place by Cys(2)His(2) (C2H2) residues in a tetrahedral array, Figure (49). C2H2 Znf's can be divided into three groups based on the number and pattern of fingers: triple-C2H2 (binds single ligand), multiple-adjacent-C2H2 (binds multiple ligands), and separated paired-C2H2. C2H2 Znf's are the most common DNA-binding motifs found in eukaryotic transcription factors, and have also been identified in prokaryotes. Transcription factors usually contain several Znf's capable of making multiple contacts along the DNA, where the C2H2 Znf motifs recognise DNA sequences by binding to the major groove of DNA via a short alpha-helix in the Znf, the Znf spanning 3-4 bases of the DNA. C2H2 Znf's can also bind to RNA and protein targets.

A particular zinc finger protein's class is determined by three-dimensional structure, but it can also be recognized based on the primary structure of the protein or the identity of the ligands coordinating the zinc ion. In spite of the large variety of these proteins, however, the vast majority typically function as interaction modules that bind DNA, RNA, proteins, or other small, useful molecules, and variations in structure serve primarily to alter the binding specificity of a particular protein.

* Vaquerizas," J." M.," Kummerfeld," S." K.," Teichmann," S." A." &" Luscombe," N." M." A" census" of" human" transcription" factors:" function," expression" and" evolution." Nature#reviews.#Genetics 10,"2527263,"doi:10.1038/nrg2538"(2009)

### 3.2.1 Why Zinc?

Zinc has four orbitals and the fourth orbital ends with two electrons. The cysteine amino acid (encoded by the codons UGU and UGC) loves to bind with zinc as the cysteine ends with the third orbit of 6 electrons, For stability, six electrons bind with 2 electrons to form common orbit of eight electrons, Figure (57). Various strategies have been developed to engineer $Cys_2His_2$ zinc fingers to bind desired sequences. These include both "modular assembly" and selection strategies that employ either Phage display or cellular selection systems.

Figure (57): Binding of $Cys_2His_2$ to DNA

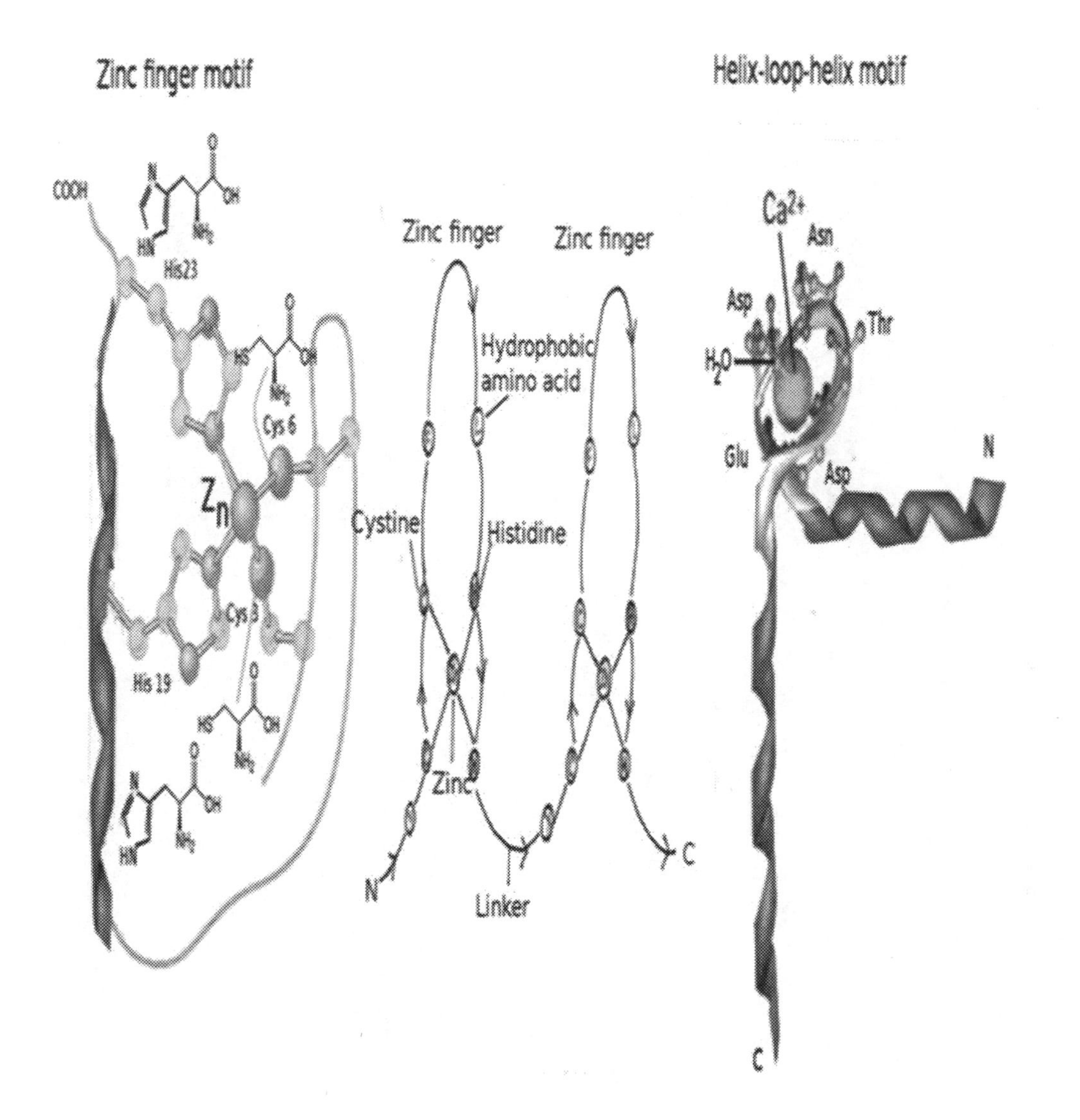


Significant understanding of DNA recognition by zinc finger came from the X-ray crystal structure of the DNA-bound Zif268 protein described by the Pabo lab. Zif268, or Egr1, is a three-finger protein that has served as the framework for understanding DNA recognition and creation of zinc fingers with novel DNA binding specificity. Zif268 was crystallized with its preferred binding site (GCGTGGGCG) as determined from prior biochemical analysis, Figure (58). The zinc finger folding, docking, and DNA recognition of Zif268 is considered as a benchmark for evaluating other zinc fingers and therefore are considered 'canonical*.

- Wolfe," S." A.," Nekludova," L." &" Pabo," C." O." DNA" recognition" by" Cys2His2" zinc" finger" proteins." Annu# Rev# Biophys# Biomol# Struct 29," 1837212," doi:10.1146/annurev.biophys.29.1.183"(2000).

Figure (58): DNA recognition by zinc finger protein

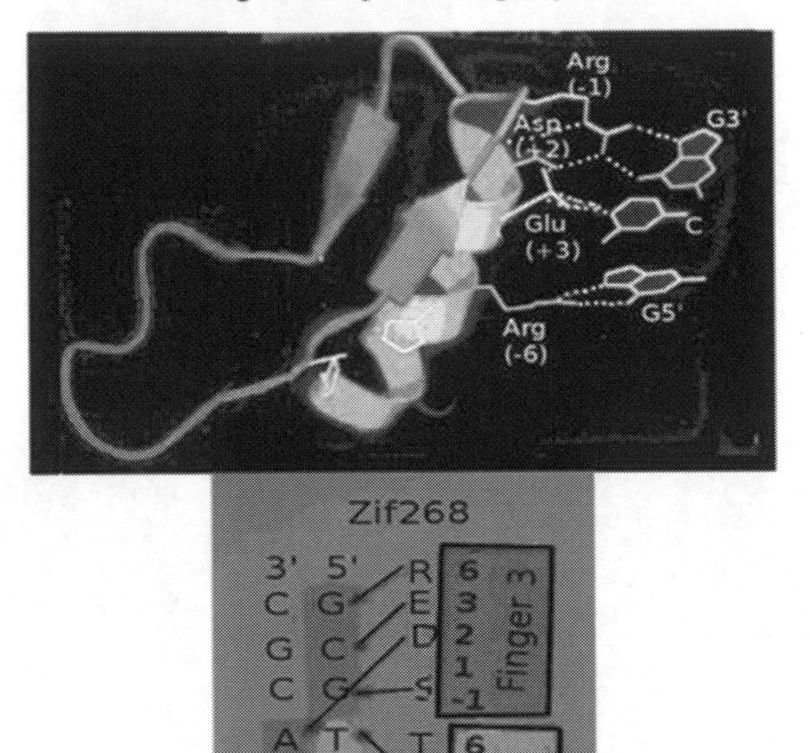

Two types of proteins (TTK and Zif) zinc C2H2 are shown in Figure (59).

Figure (59): Types of proteins of ZFN

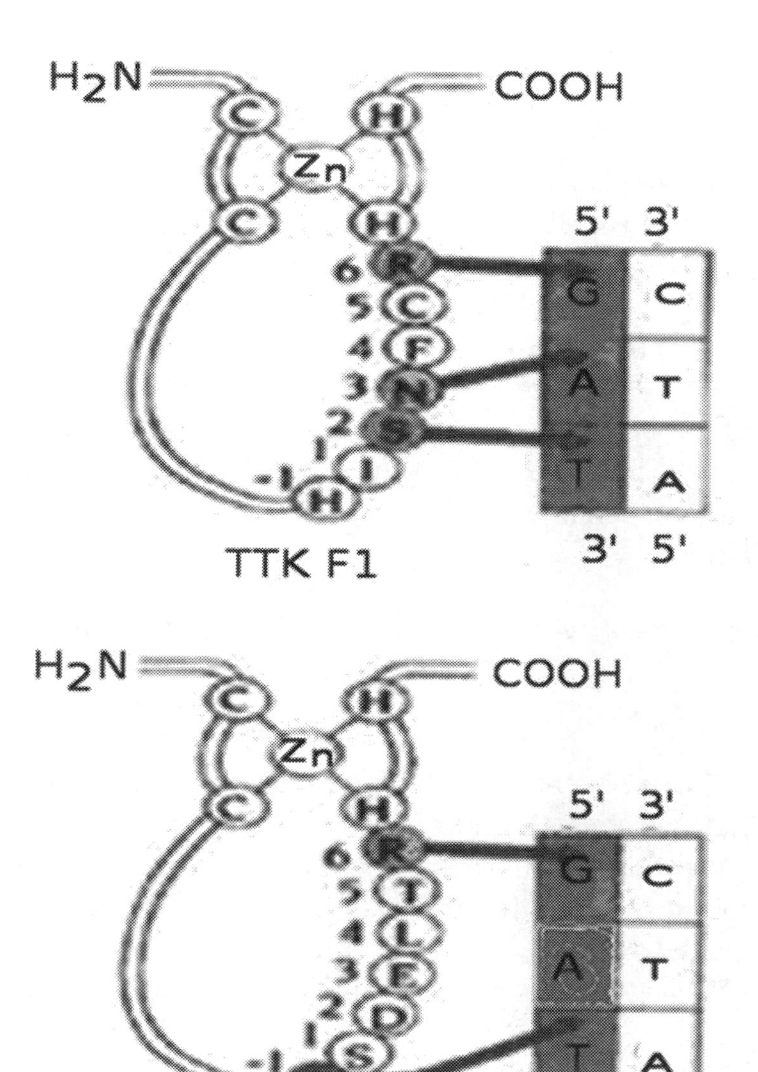


## 3.3 CRISPR/Cas system

CRISPR (Clustered regularly-interspaced short palindromic repeats) is composed from segments of prokaryotic DNA containing short repetitions of base sequences. Each repetition is followed by short segments of “spacer DNA” from previous exposures to a bacterial virus or plasmid.

CRISPR-Cas systems protect bacteria from invaders such as viruses. They do this by generating small strands of RNA that match DNA sequences specific to a given invader. When those CRISPR RNAs find a match, they unleash proteins that chop up the invader's DNA, preventing it from replicating. However, the first step in the process isn't comparing the RNA to target DNA. The first step involves PAM recognition and binding.

PAMs are short genetic sequences adjacent to the target DNA in viruses or other invaders. When the protein in a CRISPR-Cas system identifies a PAM, that identification tells the protein to bind to that DNA and begin comparing the adjacent DNA sequence to the CRISPR RNA. If the DNA and RNA match, then the protein cleaves the target DNA.

The Clustered Regularly Interspaced Short Palindromic Repeats (CRISPR) and CRISPR Associated (Cas) system was first discovered in bacteria and functions as a defense against foreign DNA, either viral or plasmid. So far three distinct bacterial CRISPR systems have been identified, termed type I, II and III. The Type II system is the basis for the current genome engineering technology available and is often simply referred to as CRISPR.

The following sequence of Type II bacterial CRISPR/Cas system is shown in

Figure (60).

2- The CRISPR array is transcribed to make the pre-CRISPR RNA (pre-crRNA).
3- The pre-crRNA is processed into individual crRNAs by a special trans-activating crRNA (tracrRNA) with homology to the short palindromic repeat. The tracrRNA helps recruit the RNAse II and Cas9 enzymes, which together separate the individual crRNAs.
4- The tracrRNA and Cas9 nuclease form a complex with each individual, unique crRNA.
5- Each crRNA (tracrRNAand Cas9 complex) seeks out the DNA sequence complimentary to the crRNA. In the Type II CRISPR system a potential target sequence is only valid if it contains a special Protospacer Adjacent Motif (PAM) directly after where the crRNA would bind.
6- After the complex binds, the Cas9 separates the double stranded DNA target and cleaves both strands after the PAM. PAMs are short genetic sequences adjacent to the target DNA in viruses or other invaders. When the protein in a CRISPR-Cas system identifies a PAM, that identification tells the protein to bind to that DNA and begin comparing the adjacent DNA sequence to the CRISPR RNA. If the DNA and RNA match, then the protein.
7- The crRNA(tracrRNA:Cas9 complex) unbinds after the double strand break.

CRISPR repeats range in size from 24 to 48 base pairs. They usually show some dyad symmetry, implying the formation of a secondary structure such as a hairpin, but are not truly palindromic. Repeats are separated by spacers of similar length. Some CRISPR spacer sequences exactly match sequences from plasmids and phages, although some spacers match the prokaryote's genome (self-targeting spacers). New spacers can be added rapidly as part of the immune response to phage infection.

Sequence of CRISPR function is shown in Figure (60).

Figure (60): Sequence of CRISPR's function

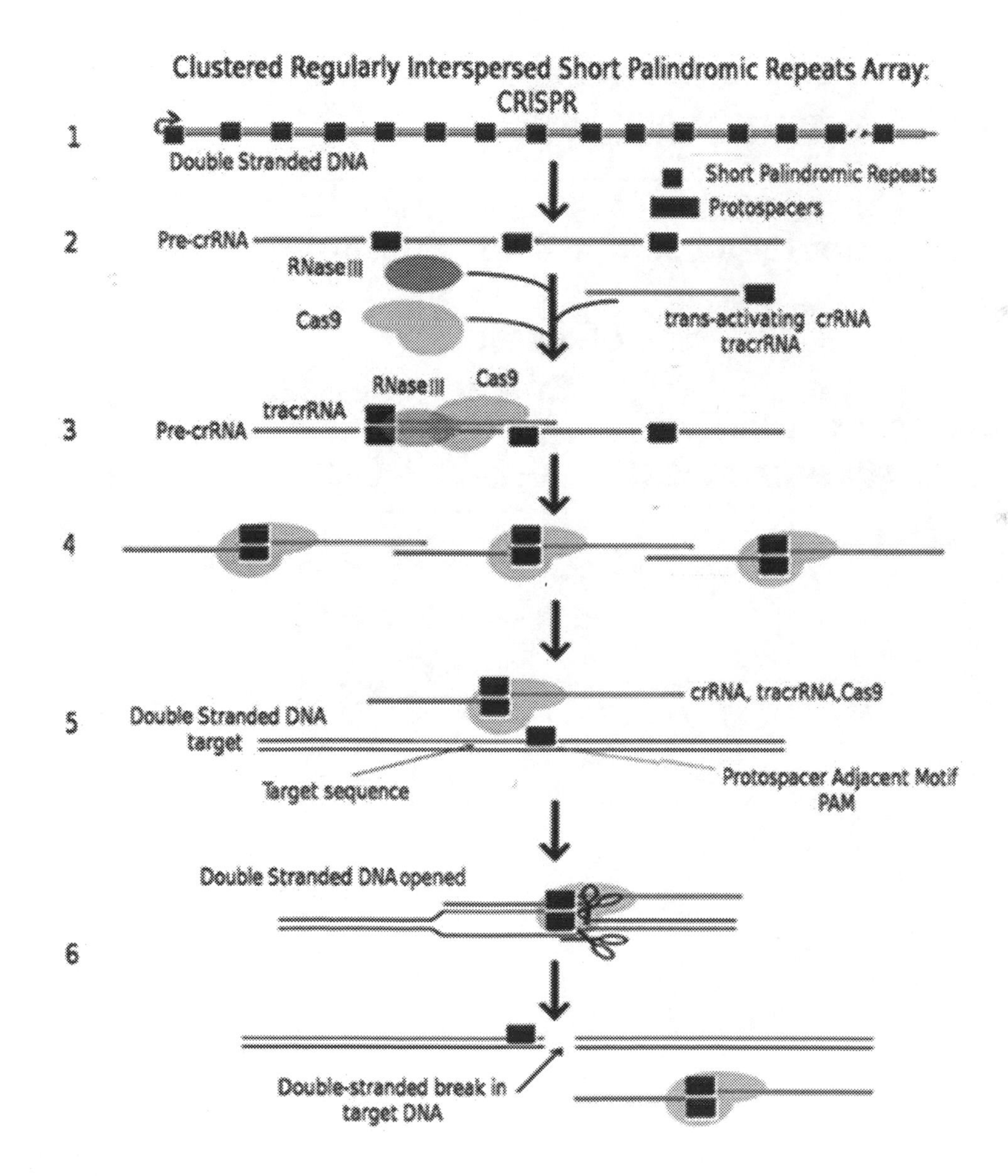


Comparision between CRISP, ZNF and TALEN is shown in Figure (61).

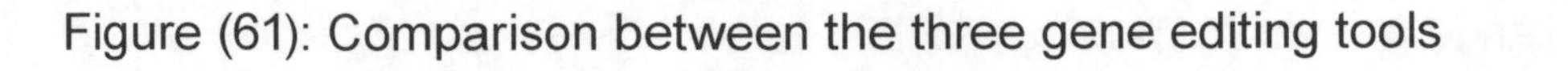

Figure (61): Comparison between the three gene editing tools

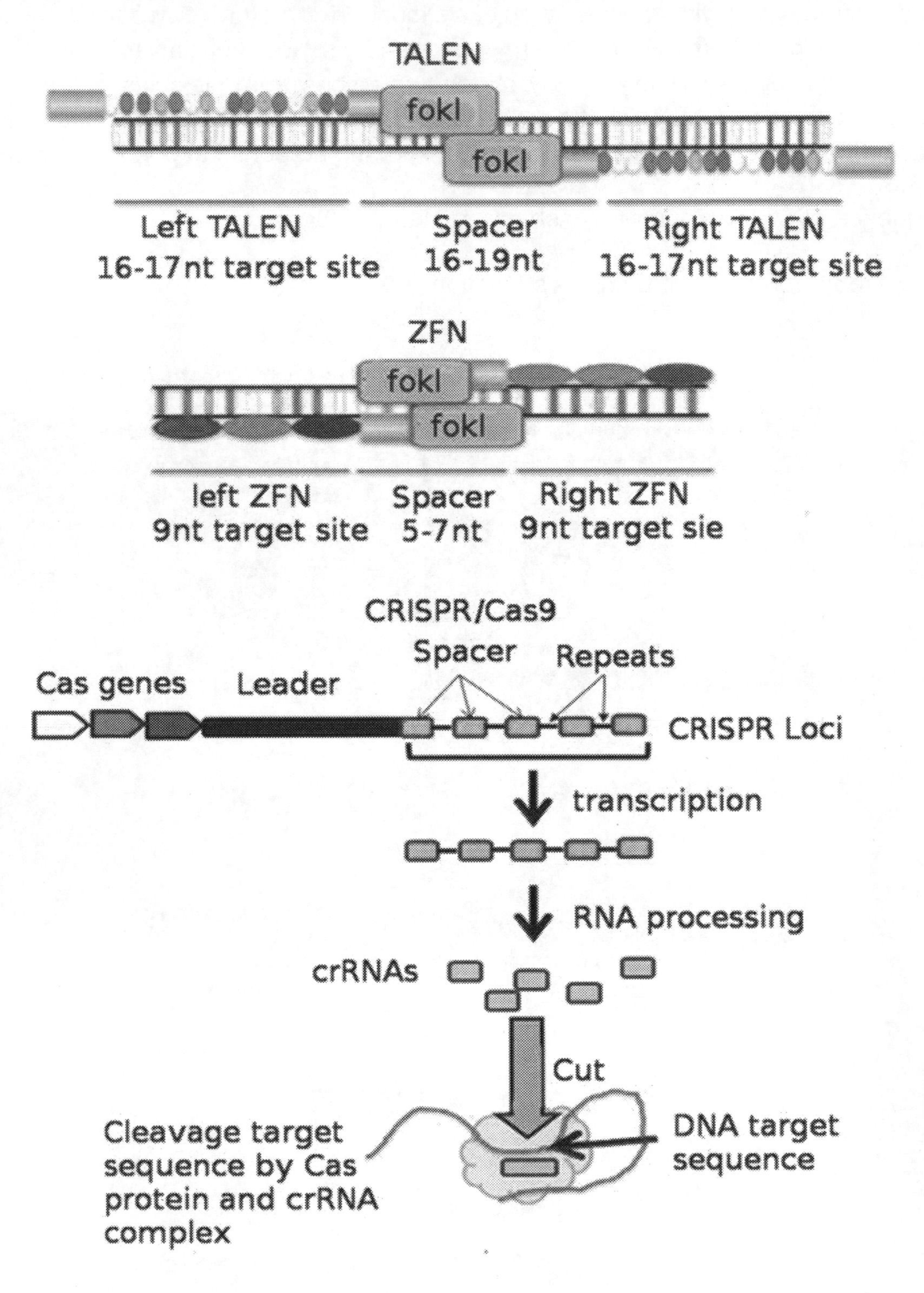

## 3.4 Genetic Manipulation

Genetic engineering, also called genetic modification, is the direct manipulation of an organism's genome using biotechnology. It is a set of technologies used to change the genetic makeup of cells, including the transfer of genes within and across species boundaries to produce improved or novel organisms.

Genetic engineering is used to:

- Growth Hormone (GH)
  - The lack of GH causes dwarfism. It used to be extracted from human corpses. Growth hormone is a 191-amino acid, single-chain polypeptide that is synthesized, stored, and secreted by somato tropic cells within the lateral wings of the anterior pituitary gland. It is produced with the help of biotechnology since 1985.
- Blood clot factor VII
  - The lack of this protein causes haemophilia. Patients used to receive regular blood transfusions which lead to possible HIV or HV infection.
- Hepatitis Virus (HV) (A & B strands) vaccines
  - This injection prevent you from suffering the illness.
- Insulin production
- Crop resistance to herbicide
- Producing a new protein or enzyme
- Introducing a novel trait

### 3.4.1 Growth Hormone Deficiency

Growth hormone deficiency is a condition initiated by a severe shortage or absence of growth hormone. Growth hormone is a protein that is necessary for the normal growth of the body's bones and tissues. Because they do not have enough of this hormone, people with isolated growth hormone deficiency commonly experience a failure to grow at the expected rate and have unusually short stature. This condition is usually apparent by early childhood.

Pituitary dwarfism is a disease caused by the lack of growth hormone (created by the gene GH1). Research shows that the condition can be treated w/ injections of human growth hormone. Growth hormone can only be obtained from human pituitary glands. These are obtained from cadavers. The studies show that the cadaver supplied growth hormones are often contaminated, so other methods need to be developed to artificially produce human growth hormone. Figure (62) shows the procedure of developing growth hormones.

Follow these steps to obtain growth hormone:

1- Isolate mRNA from cells in pituitary gland
2- Use reverse transcriptiase to synthesize a cDNA from each mRNA
3- Attach a restriction endonuclease recognition site to ends of each cDNA
4- Cut cDNA and plasmids with restriction endonucleases; remaining sticky ends join by complementary base pairing
5- Ligate cDNA and plasmids with DNA ligase
6- Bacteria cell containing gene for human growth hormone

Figure (62): Procedures of developing growth hormones

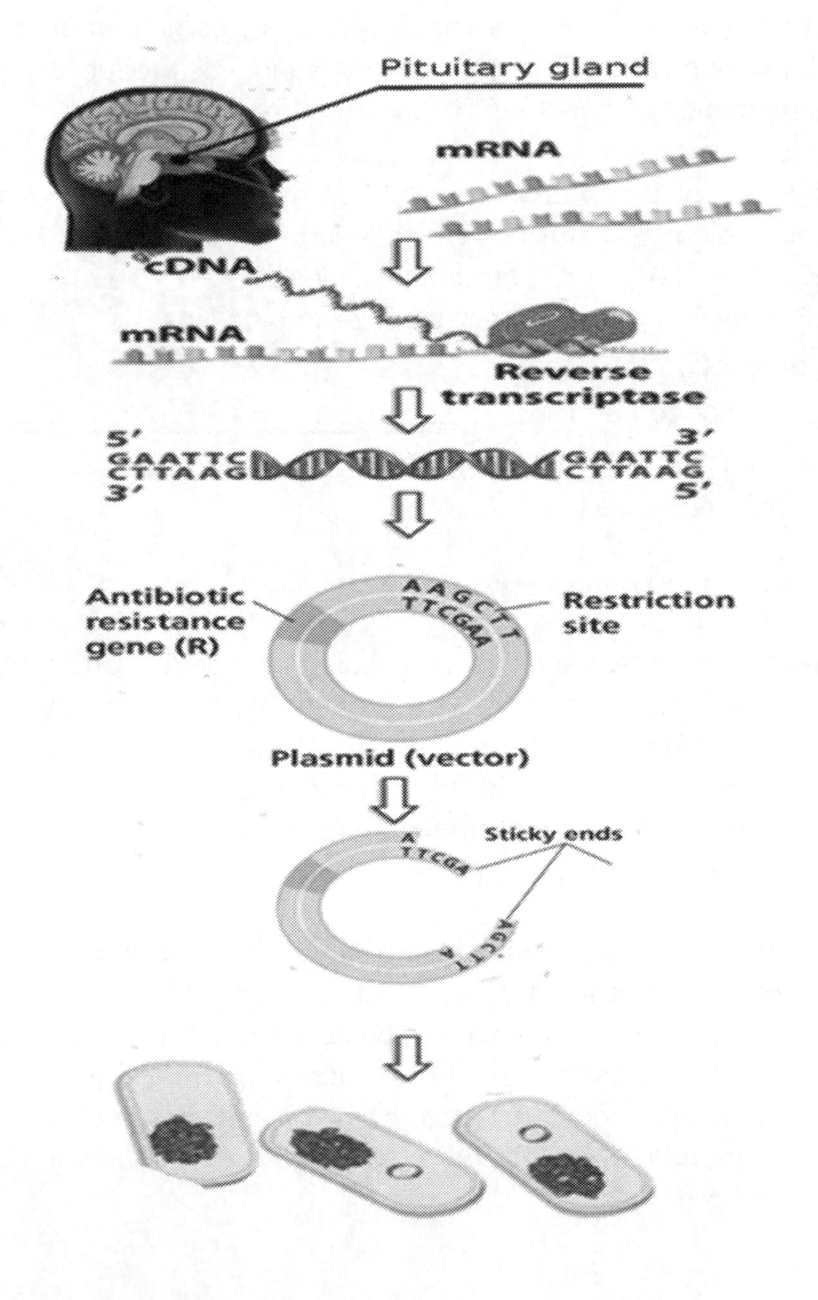

### 3.4.2 Blood clot factor VII

Factor VII, blood-coagulation factor VIIa, activated blood coagulation factor VII, is one of the proteins that causes blood to clot in the coagulation cascade. It is an enzyme of the serine protease class.. Researchers at the Children's Hospital of Philadelphia (CHOP) and elsewhere have bioengineered an adeno-associated virus (AAV), which does not cause disease, as a vector to deliver DNA into cells where it can express enough factor to make the blood clot normally. Over the past 15 years, CHOP hematology researchers have performed clinical trials of gene therapy for hemophilia B that have helped define efficacy and dosing levels in humans.

Factor VII (FVII) deficiency has a range of severity, with about 40 percent of patients having severe disease. They are most commonly treated with regular infusions of clotting factor. Unlike hemophilia, a better-known bleeding disorder that predominantly affects males, factor VII deficiency strikes males and females equally. They developed a unique animal model of this disease after identifying dogs with naturally occurring factor VII deficiency," said Margaritis. "their investigations enabled them to design the corrective gene to insert into the virus vector in the study."

### 3.4.3 Hepatitis Virus (HV) (A and B strands)

Hepatitis A (formerly known as infectious hepatitis) is an acute infectious disease of the liver caused by the hepatitis A virus (HAV). Many cases have few or no symptoms, especially in the young.

Hepatitis B virus, abbreviated HBV, is a species of the genus Orthohepadna virus, which is likewise a part of the*Hepadnaviridae* family of viruses.[1] This virus causes the disease hepatitis B.

Hepatitis A and B viruses infect liver cells *via* attachment to the cell membrane and endocytosis. The capsid with the relaxed circular (rc) DNA is released into the cytoplasm and the DNA is uncoated upon nuclear entry. In the nucleus the rcDNA is repaired to form the covalently closed circular (ccc) DNA. The cccDNA can be eventually cleared out, silenced or integrated into the host genome. In the newly assembled nucleocapsid the pgRNA serves as a template for the viral polymerase which synthesizes the rcDNA. The nucleocapsid either migrates back to the nucleus to increase the pool of cccDNA or is internalized by the endoplasmatic reticulum (ER).

Gene therapeutic strategies act on several steps of the viral replication cycle. Designer nucleases are intended to promote disruption of the cccDNA. The transcription activator-like effector (TALE) nucleases (TALEN), composed of a TALE DNA binding domain fused to the nonspecific FokI nuclease (N), are newly developed sequence-specific nucleases that could be used to introduce targeted double-stranded breaks in mammalian cells with high efficiency, while the CRISPR (clustered regularly interspaced short palindromic repeats)/Cas9 nuclease system is directed by a guide RNA (gRNA) to the target site, Figure (63).

Figure (63): Gene therapy for HBV using ZFN and CRISPR

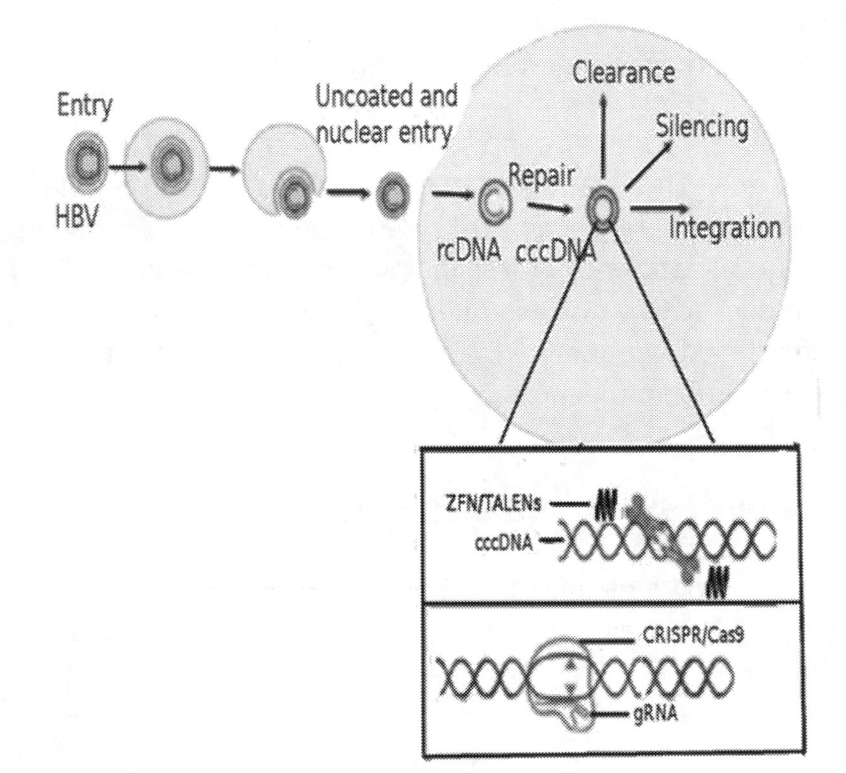


## 3.4.4 Insulin production

Natural insulin can be taken from the pancreases of pigs or cattle. However, this insulin causes adverse reactions in some people and its supply is limited.

Recent practice is to produce insulin synthetically (non naturally or man-made), using genetically modified (GM) bacteria. The gene for insulin secretion is cut from a length of human DNA and inserted into the DNA of a bacterium. The bacterium is then cultivated and quickly there are millions of bacteria producing human insulin. This genetically engineered insulin has some advantages over insulin taken from pigs or cattle:

- it can be made in very large amounts from bacteria grown in a fermenter.
- it is less likely to cause an adverse reaction
- it overcomes ethical concerns from vegetarians and some religious groups

Genetic modification needs a DNA vector and certain enzymes. Vectors take pieces of DNA and insert them into other cells. Viruses and plasmids can act as vectors.

Restriction enzymes cut DNA at specific sites, rather than just in random places along the DNA molecule. Similarly, ligase enzymes join pieces of DNA together at specific sites.

GE insulin is produced in the following way:

- The gene for making insulin is cut from a length of human DNA using restriction enzymes
- It is placed into a plasmid using ligase enzymes.
- The plasmid goes into a bacterial cell
- The transgenic bacterium reproduces, resulting in millions of identical bacteria that produce human insulin, Figure (64).

Figure (64): Insulin production

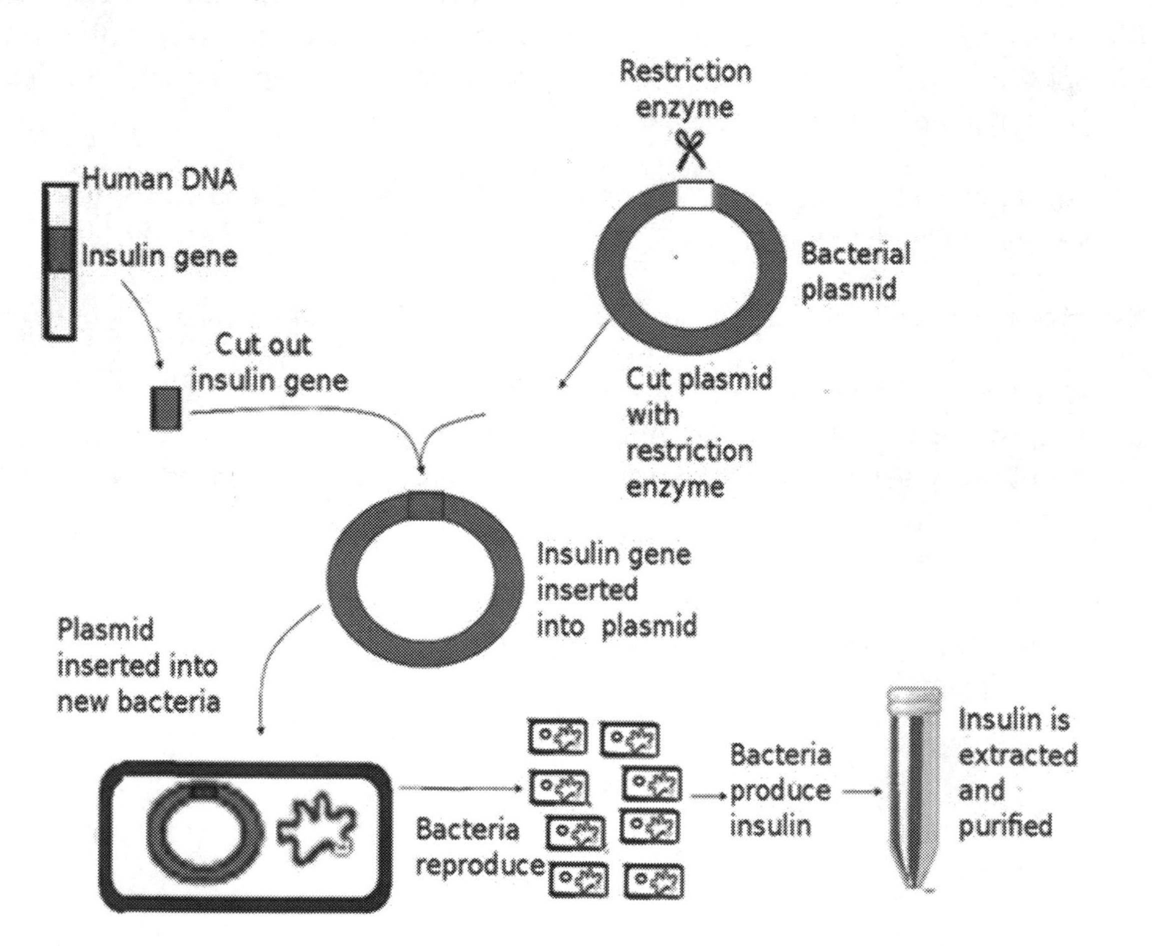

### 3.4.5 Crop resistance to herbicide

Crop management is an essential tool in modern agriculture. Herbicidal Experts in Canada have determined a process that kills the weeds growing along with the plants in a controlled manner. HTCs (Herbicide Tolerant Crops) enhance weed control options and greatly expand market demand for certain herbicides. (Knezevic, and Cassman). Modern herbicides are often synthetic mimics of natural plant hormones which interfere with growth of the target plants. The term organic herbicide has come to mean herbicides intended for organic farming; these are often less efficient and more costly than synthetic herbicides and are based upon natural materials

Genetically modified crops (GMCs, GM crops, or biotech crops) are plants used in agriculture, the DNA of which has been modified using genetic engineering techniques. In most cases, the aim is to introduce a new trait to the plant which does not occur naturally in the species. Examples in food crops include resistance to certain pests, diseases, or environmental conditions, reduction of spoilage, or resistance to chemical treatments (e.g. resistance to a herbicide), or improving the nutrient profile of the crop. Examples in non-food crops include production of pharmaceutical agents, biofuels, and other industrially useful goods, as well as for bioremedation.

Farmers have widely adopted GM technology. Between 1996 and 2013, the total surface area of land cultivated with GM crops increased by a factor of 100.

The following method can be used to produce crop resistance to herbicide, Figure (65).

- Isolate the herbicide resistance gene from the Salmonella
- Splice it into a plasmid from Agrobacterium
- Put the plasmid and the herbicidal gene (from above) into agrobacterium tumefaciens
- Culture them
- Transfer the leaf to the culture
- Transfer a growing tip to new growing medium
- transplant

Figure (65): Production of crop resistance to herbicide

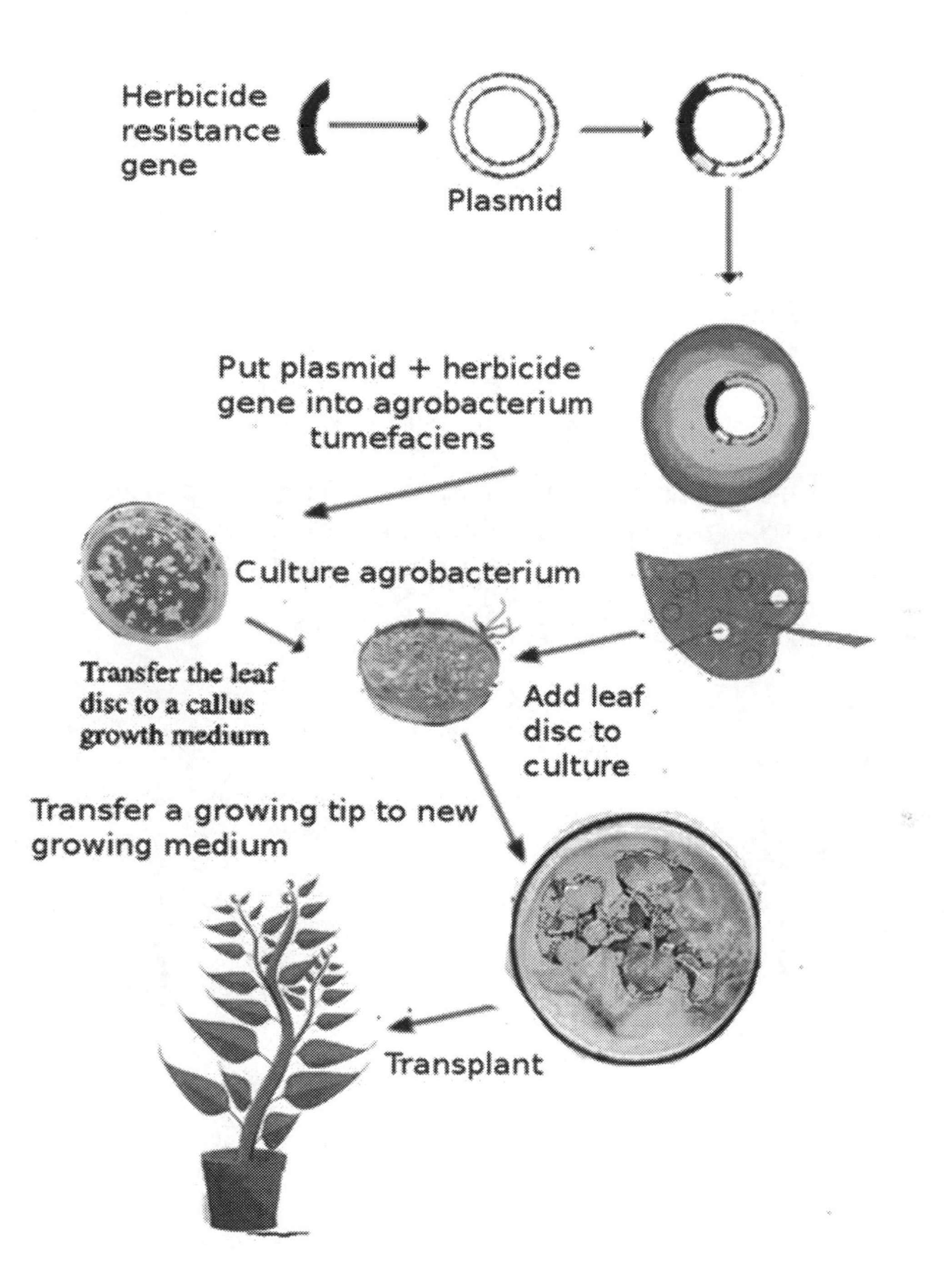


### 3.4.6 Producing a new protein or enzyme

*Eco*RI (pronounced, "eco R one") is a restriction endonuclease enzyme isolated from species E. coli*.* The *Eco* part of the enzyme's name originates from the species from which it was isolated, while the R represents the particular strain, in this case RY13. The last part of it's name, the I, denotes that it was the first enzyme isolated from this strain. *Eco*RI is a restriction enzyme that cleaves DNA double helixes into fragments at specific sites. It is also a part of the restriction modification system.

The traditional method for creating recombinant DNA typically involves the use of plasmids in the host bacteria. The plasmid contains a genetic sequence corresponding to the recognition site of a restriction endonuclease, such as EcoR! (EcoR1 is an endonuclease enzyme isolated from strains of E.coli). After foreign DNA fragments, which have also been cut with the same restriction endonuclease, have been inserted into host cell, the restriction endonuclease gene is expressed by applying heat, or by introducing a biomolecule, such as arabinose. Upon expression, the enzyme will cleave the plasmid at its corresponding recognition site creating sticky ends on the plasmid. Ligases then joins the sticky ends to the corresponding sticky ends of the foreign DNA fragments creating a recombinant DNA plasmid, Figure (66).

Advances in genetic engineering have made the modification of genes in microbes quite efficient allowing constructs to be made in about a weeks worth of time. It has also made it possible to modify the organism's genome itself.

Figure (66): production of protein or enzyme

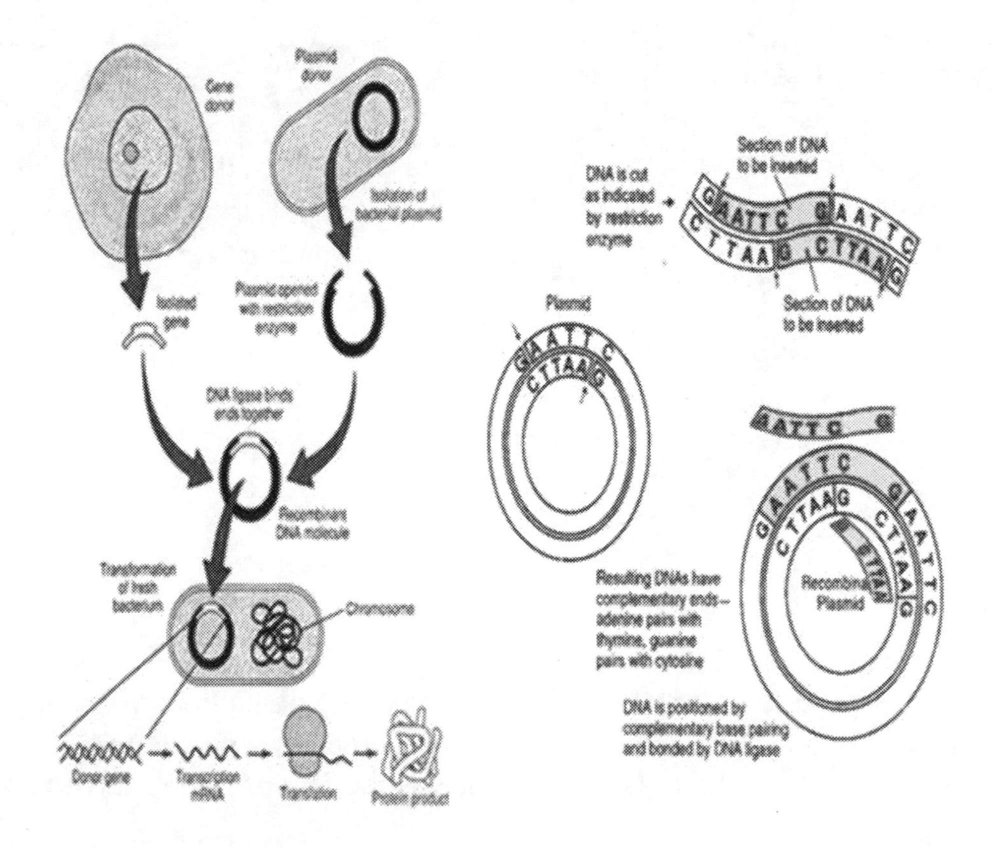

### 3.4.7 Introducing a novel trait (Canada)

In common language, “novelty” refers to something previously unknown. The Canadian Food Inspection Agency (CFIA) uses this meaning when applying the word “novelty” specifically to some of the products it regulates, such as plants, livestock feeds, and plant and soil supplements. This fact sheet focuses on plants with novel traits.

The CFIA defines a plant with a novel trait (PNT) as a new variety of a species that has one or more traits that are novel to that species in Canada. A trait is considered to be novel when it has both of these characteristics:

- it is new to stable, cultivated populations of the plant species in Canada, and
- it has the potential to have an environmental effect.

These PNTs are assessed for safety for the environment. For more details see the CFIA's Directive 94-08 (Dir94-08) Assessment Criteria fo Determining Environment Safety of Plants with Novel Traitss.

To date, in Canada, all genetically engineered plants have been considered to contain novel traits and, therefore, have been assessed for safety.

However, the approach used by the CFIA does not mean that all PNTs are developed through genetic engineering. Novel traits can be developed through various techniques. Examples (other than genetic engineering) are mutagenesis, cell fusion, and traditional breeding. For more information, see the factsheet “Modern Biochnology: A brief Overview”.

During extensive consultations, Canadian scientists and other stakeholders recognized that the potential for risk lies with the new trait and not with the process by which the trait was introduced. For example, techniques that are not considered genetic engineering, such as mutagenesis, could lead to a product that is novel from a safety perspective.

When regulating plant products of biotechnology, no other country uses as broad a regulatory scope as Canada does with its “novelty” approach.

# CHAPTER 4

# CLONING

In biology, cloning is the process of producing similar populations of genetically identical individuals that occurs in nature when organisms such as bacteria, insects or plants reproduce asexually. Cloning in biotechnology refers to processes used to create copies of DNA fragments (molecular cloning), cells (cell cloning), or organisms. The term also refers to the production of multiple copies of a product such as digital media or software.

Lots of people first heard of cloning when Dolly the Sheep showed up on the scene in 1997. Artificial cloning technologies have been around for much longer than Dolly, though.

The term cloning describes a number of different processes that can be used to create genetically identical copies of a biological entity. The copied material, which has the same genetic makeup as the original, is referred to as a clone. Researchers have cloned a wide range of biological materials, including genes, cells, tissues and even full organisms, such as a sheep.

There are two ways to make an exact genetic copy of an organism in a lab: artificial embryo twinning and somatic cell nuclear transfer.

## 4.1 Artificial embryo twinning

Artificial twinning is also called embryo cloning or embryo splitting. Identical twins can form naturally when an early embryo splits and two foetuses grow from the two resulting embryos: artificial twinning occurs in the same way.

In nature, twins form very early in development when the embryo splits in two. Twinning happens in the first days after egg and sperm join, while the embryo is made of just a small number of unspecialized cells. Each half of the embryo continues dividing on its own, ultimately developing into separate, full and complete individuals. Since they developed from the same fertilized egg, the resulting individuals are genetically identical.

Artificial embryo twinning makes use of the same approach, but it is carried out in a Petri dish instead of inside the mother. Research has found that unfertilized eggs can b activated either by (1) an electrical jolt, by (2) chemical stimulation or by (3) manual. A very early embryo is split into individual cells, which are allowed to divide and develop for a short time in the Petri dish, Figure (67). The embryos are then placed into a surrogate (substitute) mother, where they finish developing. Again, since all the embryos came from the same fertilized egg, they are genetically identical.

Figure (67): Artificial Embryo twinning

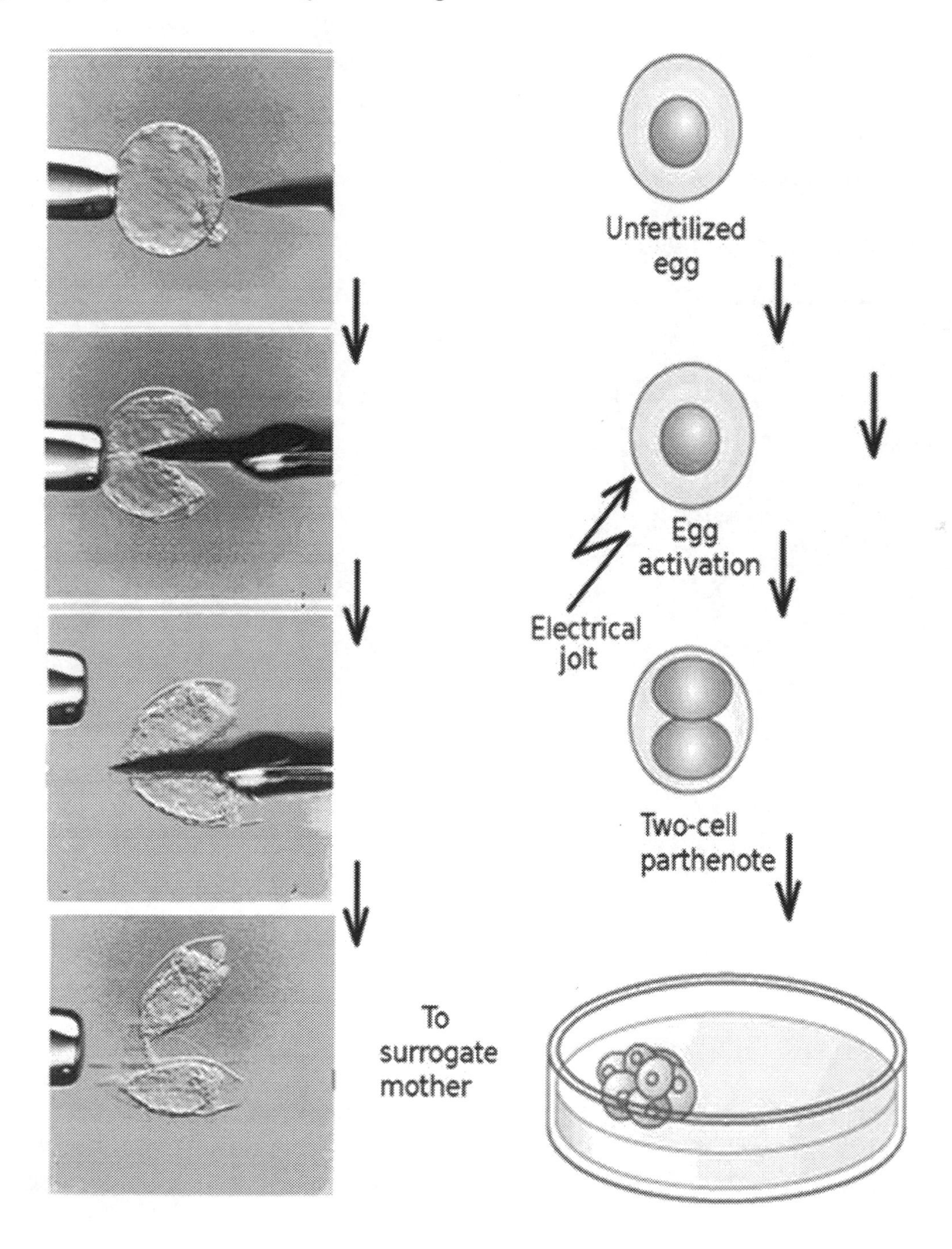

## 4.2 Somatic cell nuclear transfer

Applying Somatic Cell Nuclear Transfer (SCNT) in the creation of Dolly the cloned sheep was the first mammal to have been successfully cloned from an adult cell. She was cloned at the Roslin Institute in Midlothian, Scotland, and lived there until her death when she was six years old. Her birth was announced on February 22, 1997. Dolly the sheep was cloned through somatic cell nuclear transfer (SCNT). An adult cell from the mammary gland of a Finn-Dorset ewe acted as the nuclear donor; it was fused with an enucleated egg from a Scottish Blackface ewe, which acted as the cytoplasmic (or egg) donor. An electrical pulse acted to fuse the cells and activate the oocyte after injection into the surrogate mother ewe. A successfully implanted oocyte developed into the lamb Dolly, a clone of the nuclear donor, the Finn-Dorset ewe. SCNT may also be used to create patient-specific stem cells with great therapeutic potential, Figure (68).

Figure (68): Cloning of somatic cells

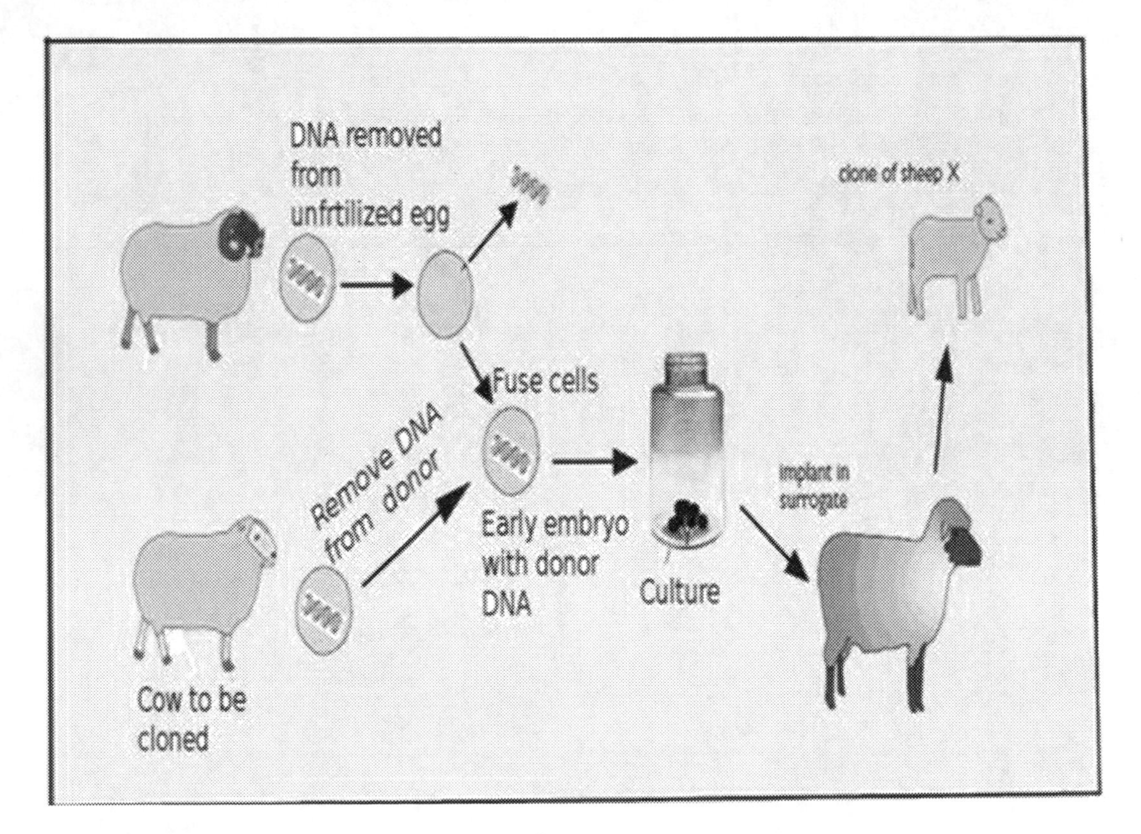


## 4.3 Asexual reproduction

Asexual reproduction is a type of reproduction by which offspring develop from a single organism, and inherit the genes of that parent only; it does not involve the fusion of gametes and almost never changes the number of chromosomes. Asexual reproduction is the primary form of reproduction for single-celled organisms such as the archaebacteria, eubacteria, and protists. Many plants and fungi reproduce asexually as well.

There are six types of asexual reproduction:

### 4.3.1 Binary Fission

In binary fission, the parent organism is replaced by two daughter organisms, because it literally divides in two. Organisms, both prokaryotes (the archea and the bacteria), and eukaryotes (such as protists and unicellular fungi), reproduce asexually through binary fission; most of these are also capable of sexual reproduction, Figure (69).

Another type of fission is multiple fission that is advantageous to the plant life cycle. Multiple fission at the cellular level occurs in many protists, e.g. sporozoans and algae. The nucleus of the parent cell divides several times by mitosis, producing several nuclei. The cytoplasm then separates, creating multiple daughter cells,.

Figure (69): Binary fission

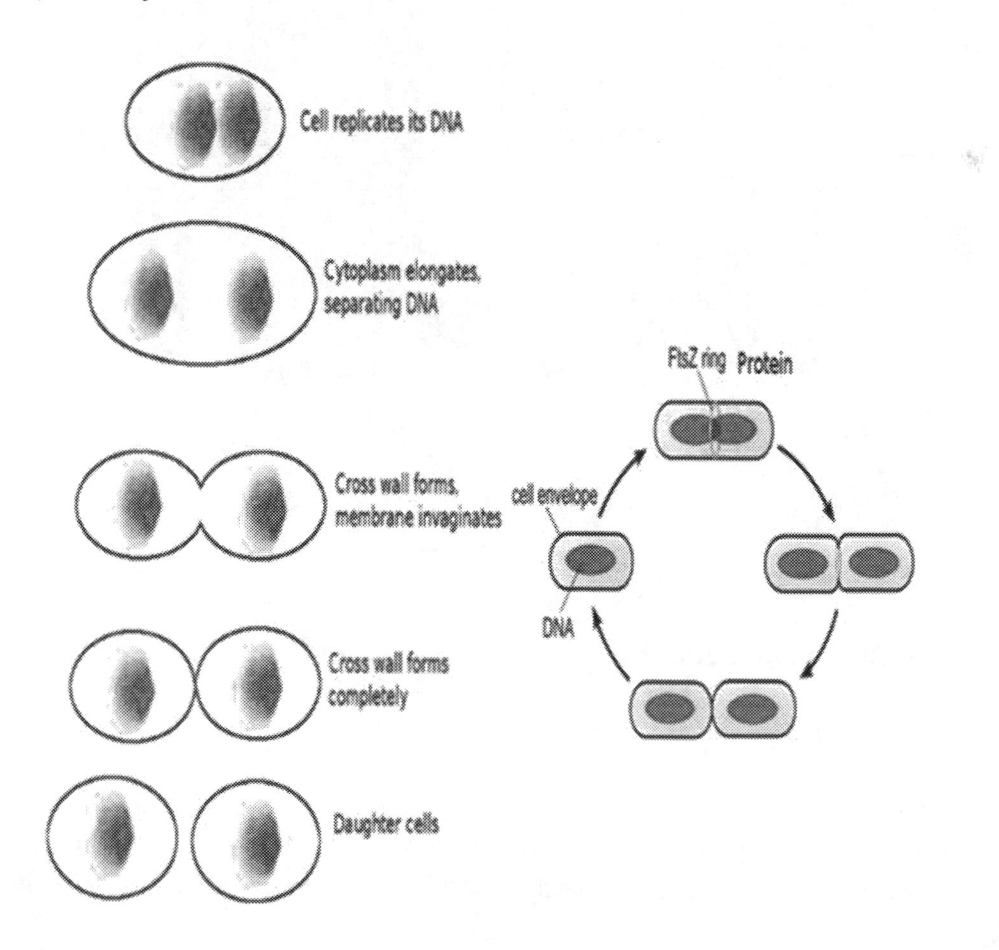

### 4.3.2 Budding

Budding is a form of asexual reproduction in which a new organism develops from an outgrowth or bud due to cell division at one particular site.

Some cells split via budding (for example baker's yeast), resulting in a 'mother' and 'daughter' cell. The offspring organism is smaller than the parent. Budding is also known on a multicellular level; an animal example is the hydra, which reproduces by budding. The buds grow into fully matured individuals which ultimately break away from the parent organism.

In this type of asexual reproduction, bulblike projections called buds arise from the parent body. Mature yeast cells are larger, and spherical or oval in shape. One or more bulblike projections (buds) develops from the cell membrane. The nucleus of the parent cell divides and one of the daughter nuclei passes into the bud. The bud is finally separated from the parent body and grows into a new individual. The parent's identity is maintained in budding, Figure (70).

Figure (70): Budding

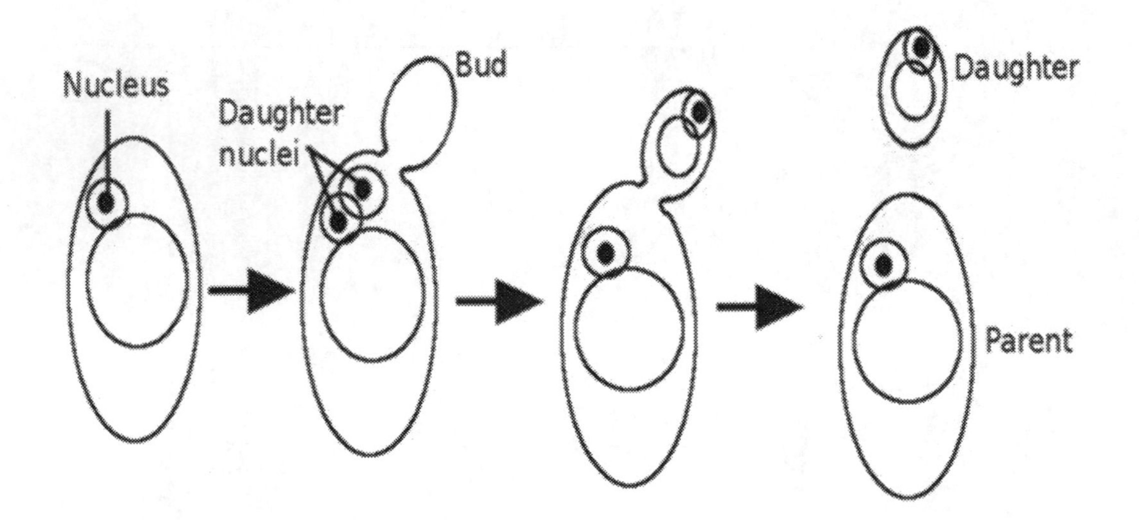


### 4.3.3 Vegetative Reproduction

Mainly, there are three types of vegetative reproductions:

1- Grafting: In grafting 2 plants are used to develop a new plant with blended traits from the 2 parent plants. In grafting the scion is the above ground part of one plant. The scion is attached to the stock which is the rooted part of the second plant.

2- Layering: In layering a shoot of a parent plant is bent until it can be covered by soil. The tip of the shoot stays above ground. New roots and eventually a new plant will grow up. These plants can then be separated.
3- Stems: Runners are stems that grow horizontally above the ground. They have nodes where buds are produced. These buds grow into a new plant, Figure (71).

Figure (71): Vegetative production

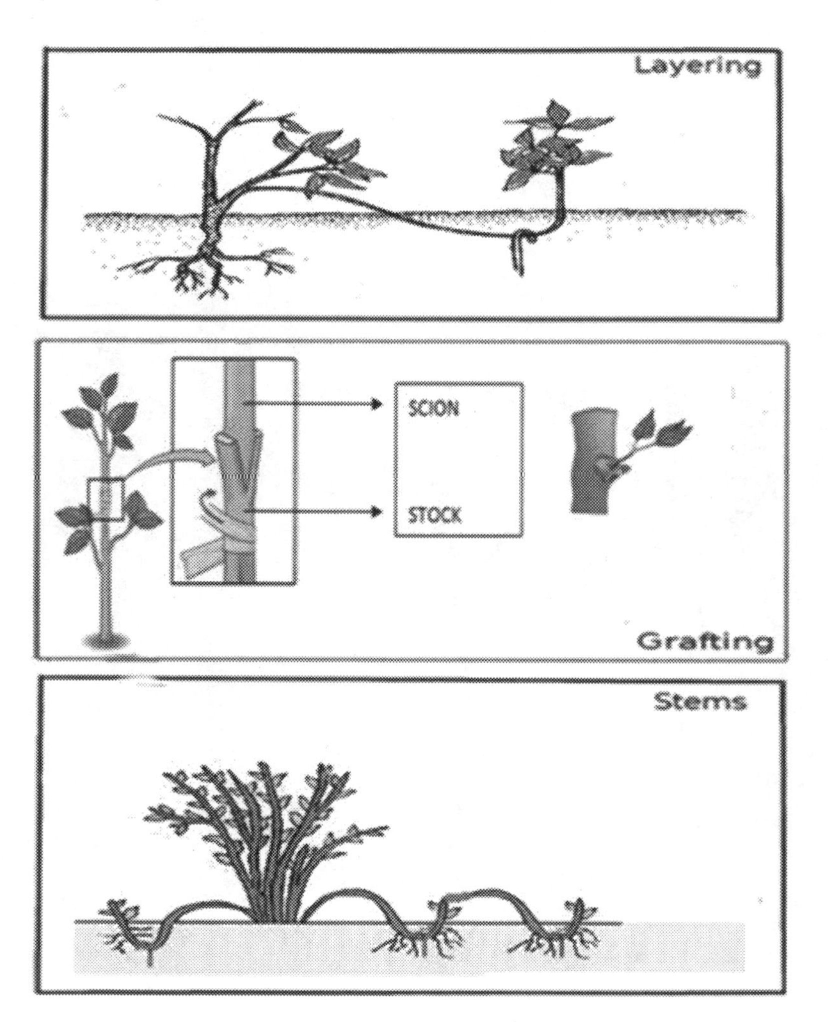


### 4.3.4 Spore

The fungi on a bread piece grow from spores which are there in the air. When spores are liberated they keep floating in the air. As they are very light they can cover long distances. Fungi and plants are sessile (immobile). Unlike animals, they cannot walk or fly to new habitats. Their immobility generally leaves only two ways for fungi and plants to extend their range: they can grow into an adjoining area, or disperse spores or seeds. Most fungal spores are single cells. They can travel beyond the physical limits of their parent into more distant territory.

The spores are asexual reproductive bodies. Each spore is covered by a hard protective coat to withstand unfavorable environment conditions such as high temperature and low humidity. So they can survive for a long time. Under favorable conditions, a spore germinates and grows into a new individual. Plants, Figure (72).

Spores form part of the life cycles of many plants, algae, fungi and protozoa. Bacterial spores are not part of asexual cycle but are resistant structures used for survival under unfavorable conditions. Myxozoan spores release amoebulae into their hosts for parasitic infection, but also reproduce within the hosts through the pairing of two nuclei within the plasmodium, which develops from the amoebula.

Figure (72): Spores of fungi are asexual reproductive bodies

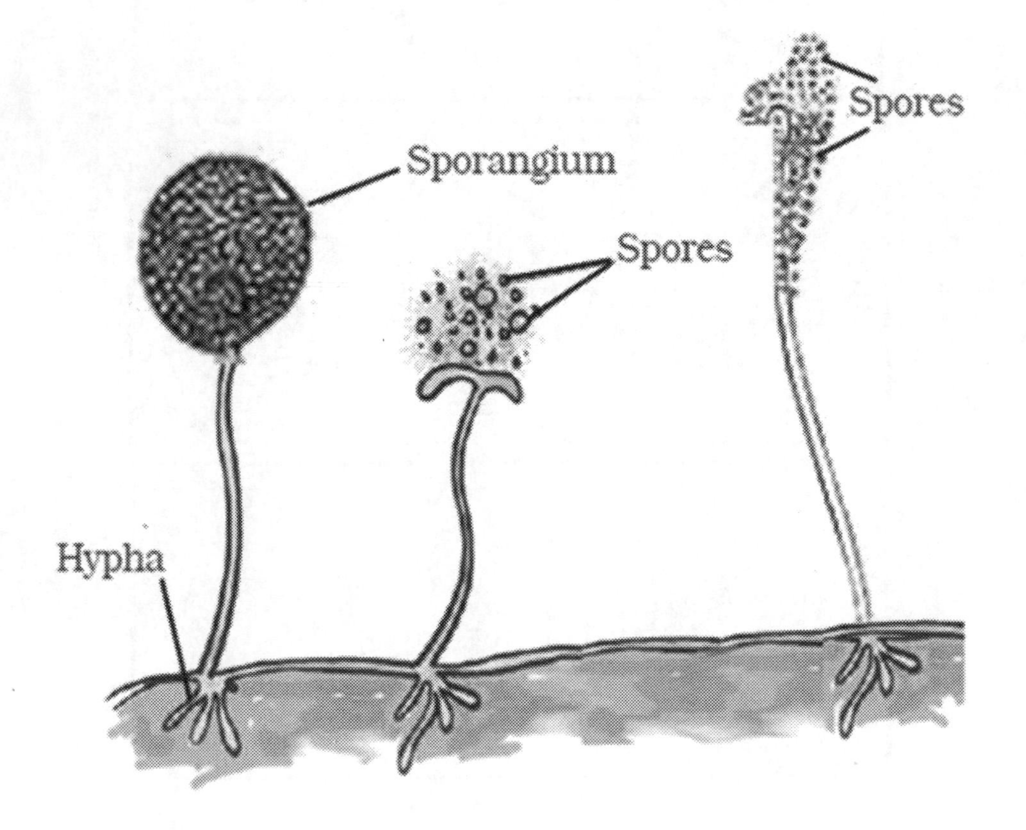


### 4.3.5 Fragmentation

Fragmentation or clonal fragmentation in multicellular or colonial organisms is a form of asexual reproduction or cloning in which an organism is split into fragments. Fragmentation may be categorized as the process of breaking up of parent animal into small parts, each of which can grow into a new complete individual. This process of asexual reproduction is found in planaria and hydra, Figure (73).

Figure (73): Reproduction by fragmentation

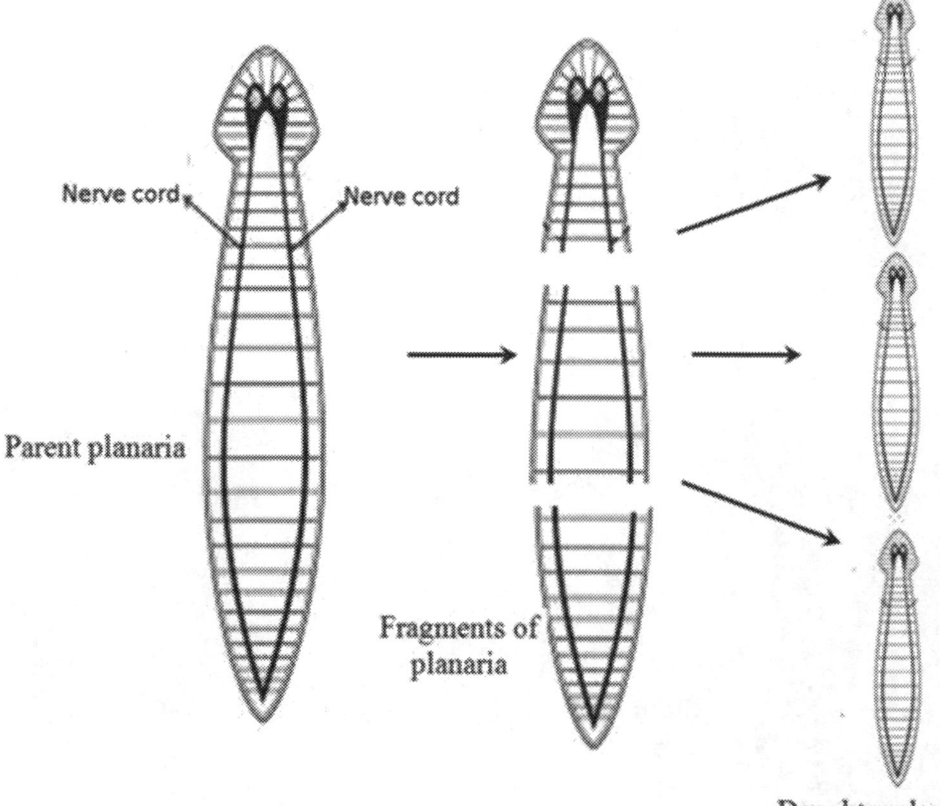


### 4.3.6 Agamogenesis

Agamogenesis is any form of reproduction that does not involve a male gamete. Examples are parthenogenesis and apomixis. Agamogenesis is any shape of reproduction that does not include a male gamete. Examples are parthenogenesis and apomixis. Asexual reproduction by special types of New Zealand mud snails is through agamogenesis reproduction, Figure (74).

Figure (74): Asexual reproduction by agamogenesis

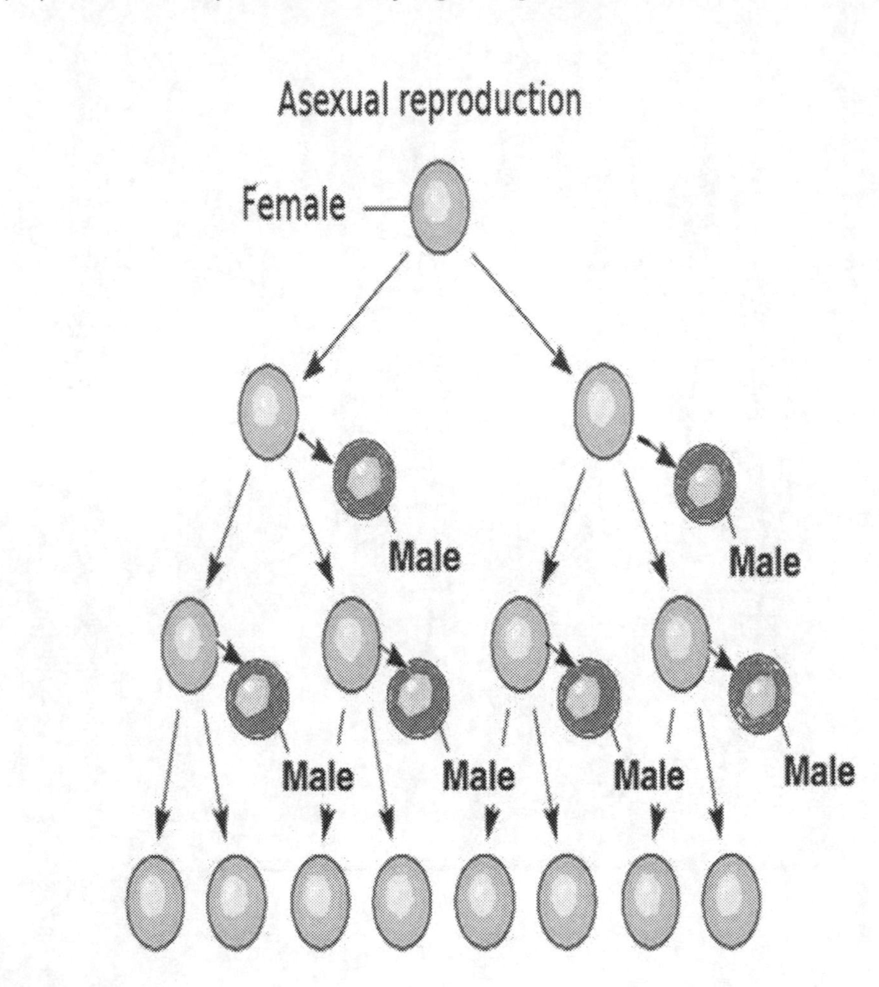


## 4.4 Alternation between Sexual and Asexual Reproduction

Asexual reproduction is a type of reproduction by which offspring arise from a single organism, and inherit the genes of that parent only; it does not involve the fusion of gametes and almost never changes the number of chromosomes. Asexual reproduction is the primary form of reproduction for single-celled organism as the archaebacteria, eubacteria, and protists. Many plants and fungi reproduce asexually as well. Some species interchange between the sexual and asexual strategies, an ability known as heterogamy, depending on conditions. Alternation is observed in several rotifier species (cyclical parthenogenesis e.g. in Branchionus species) and a few types of insects, such as aphids which will, under certain conditions, produce eggs that have not gone through meiosis, thus cloning themselves. The cape bee Apis mellifera can replicate asexually through a process called the lytoky. A few species of amphibians, reptiles, and birds have a similar ability (see parthenogenesis for examples). For example, the freshwater crustacean Daphnia reproduces by parthenogenesis in the spring to quickly populate ponds, and then switches to sexual reproduction as the intensity of competition and predation increases. Another example are monogonont rotifers of the genus Brachionus, which reproduce via cyclical parthenogenesis: at low population densities females produce asexually and at high densities a chemical cue accumulates and induces the transition to sexual reproduction. Many protists and fungi alternate between sexual and asexual reproduction, Figure (75).

For Instance, the slime mold Dictovostelium undergoes binary fission (mitosis) as single-celled amoebae under favorable conditions. However, when conditions turn unfavorable, the cells aggregate and follow one of two different developmental pathways, depending on conditions. In the social pathway, they grow in a multicellular slug which then forms a fruiting body with asexually generated spores. In the sexual pathway, two cells fuse to form a giant cell that develops into a large cyst. When this macrocyst germinates, it releases hundreds of amoebic cells that are the product of meiotic recombination between the original two cells.

The hyphae of the common mold (Rhizopus) are capable of producing both mitotic as well as meiotic spores. Many algae similarly switch between sexual and asexual reproduction. A number of plants use both sexual and asexual means to produce new plants; some species alter their primary modes of reproduction from sexual to asexual under varying environmental conditions.

Figure (75): Alternation between Sexual and Asexual Reproduction

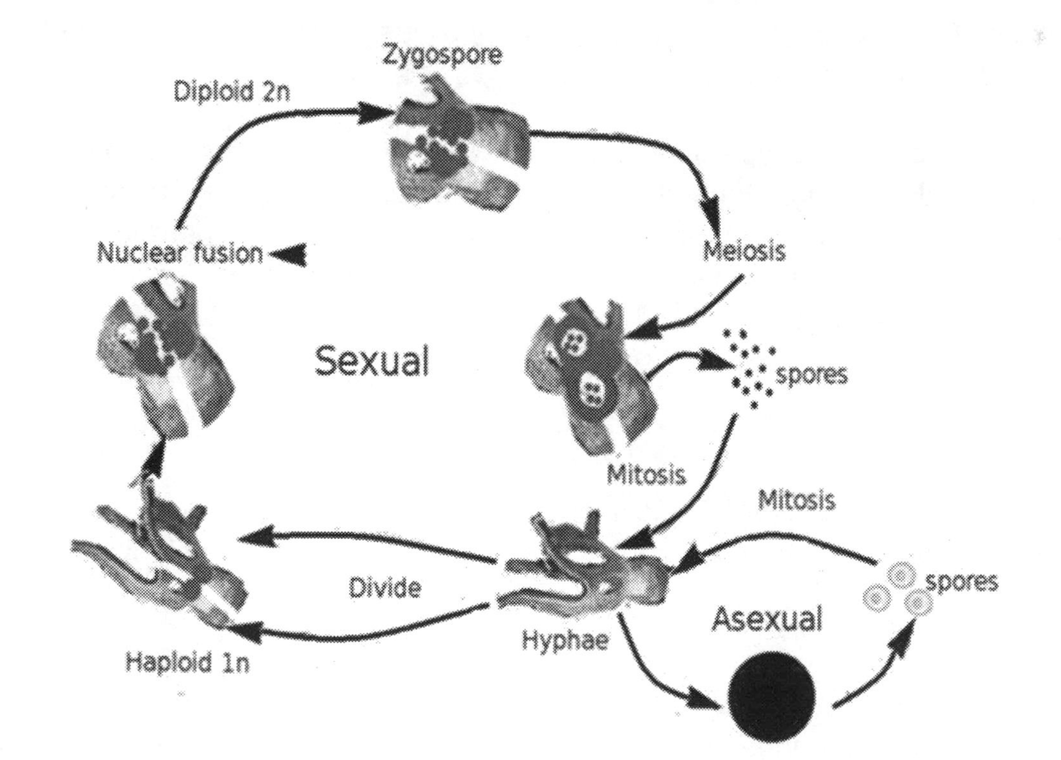

## 4.5 Inheritance of asexual reproduction in sexual species

In some species, for example, in the rotifer Brachionus calvifiorus asexual reproduction (obligate parthenogenesis) can be inherited by a recessive allele, which leads to loss of sexual reproduction in homozygous offspring, Scheuerl, T.; Riss, S.; Stelzer, C.P. (2011). "Phenotypic effects of an allele causing obligate parthenogenesis in a rotifer". Journal of Heredity 102 (4): 409–415.

Inheritance of asexual reproduction by a single recessive locus has also been found in the parasitoid wasp Lysiphlebus fabarum.

## 4.6 Cloning

Cloning is the process of producing similar populations of genetically alike individuals that occurs in nature when organisms such as bacteria, insects or plants reproduce asexually. The copied material, which has the same genetic makeup as the original, is referred to as a clone. Researchers have cloned a varied range of biological materials, including genes, cells, tissues and even entire organisms, such as a sheep.

Natural cloning is a natural form of reproduction that has allowed life forms to spread for more than 50 thousand years. It is the reproduction method used by plants, fungi and bacteria, and is also the way that cloned colonies reproduce themselves. Examples of these organisms include blueberry plants, hazel trees, the Pando trees, the Kentucky coffee trees.

There are three different types of artificial cloning: gene cloning, reproductive cloning and therapeutic cloning. Gene cloning produces copies of genes or segments of DNA. Reproductive cloning produces copies of whole animals. Therapeutic cloning produces embryonic stem cells for experiments aimed at creating tissues to replace injured or diseased tissues.

### 4.6.1 Artificial Cloning

There are many methods to clone animals artificially: Here. We give four main methods: Embryo transplants, Fusion cell cloning, Natural cloning and Gene cloning.

- Embryo transplants

A developing embryo is detached from a pregnant animal at an early stage, before its cells have had time to become specialized. The cells are separated, grown for a while in a laboratory then transplanted into substitute mothers.

When the offspring are born, they are identical to each other and to the original pregnant animal. They are not identical to their host mothers because they contain different genetic information, Figure (76).

Figure (76): Transplant of embryos

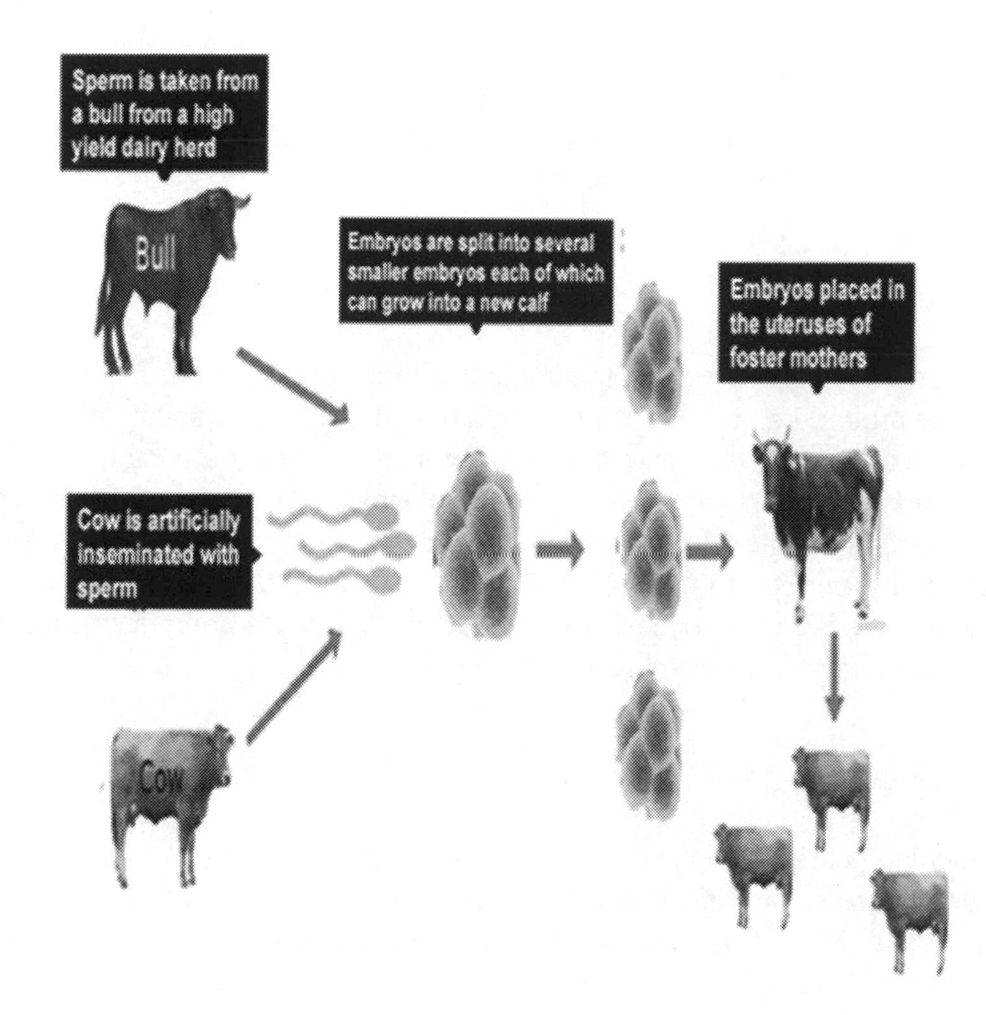


## 4.6.2 Fusion Cell Cloning

This has been argued in earlier section of this book. Fusion cell cloning involves replacing the nucleus of an unfertilised egg with one from a different cell. The replacement can come from an embryo. 'Dolly the sheep' was the first mammal to be cloned using adult cell cloning. Here is how she was produced: It is repeated here.

1. An egg cell was removed from the ovary of an adult female sheep, and its nucleus removed.
2. The nucleus from an udder cell of a donor sheep was inserted into the empty egg cell.

3. The fused cell then began to develop normally, using genetic information from the donated DNA.
4. Before the dividing cells became specialized, the embryo was implanted into the uterus of a foster mother sheep. The result was Dolly, who was genetically identical to the donor sheep.

### 4.6.3 Natural Cloning

Twins are genetically identical because they are formed after one fertilized egg cell splits into two cells. They are natural clones.

## 4.7 Gene Cloning

Researchers regularly use cloning techniques to make copies of genes that they wish to study. The procedure consists of inserting a gene from one organism, often referred to as "foreign DNA," into the genetic material of a carrier called a vector. Examples of vectors include bacteria, yeast cells, viruses or plasmids, which are small DNA circles carried by bacteria. After the gene is placed in, the vector is placed in laboratory conditions that prompt it to multiply, resulting in the gene being copied many times over. The process is similar to the insulin production which has been discussed earlier. It is reproduced here in different way, Figure (77).

### 4.7.1 Human Cloning

Human cloning is the creation of a genetically identical copy of a human. The term is usually used to refer to artificial human cloning, which is the reproduction of human cells and tissue. It does not refer to the natural conception and delivery of identical twins. The possibility of human cloning has raised controversies debates. These ethical concerns have prompted several countries to pass laws regarding human cloning and its legality.

Two commonly discussed types of theoretical human cloning are: therapeutic cloning and reproductive cloning. Therapeutic cloning would involve cloning cells from a human for use in medicine and transplants, and is an active area of research, but is not in medical practice anywhere in the world, as of January 2016. Two common methods of therapeutic cloning that are being researched are somatic-cell nuclear transfer and, more recently, pluripotent stem cell induction. Reproductive cloning would involve making an entire cloned human, instead of just specific cells or tissues,

Most of cloned mammals experience abnormalities in number of chromosomes. The effect of chromosome abnormalities, classified into three main categories, is not uniform. Euploidies (having a balanced set or sets of chromosomes, 2, 3, or any number), autosomal aneuploidies (any variation in chromosome number (usually less than or more than 46) that involves individual chromosomes rather than entire sets of chromosomes and large unbalanced structural rearrangements usually cause early embryonic death. An exception is found in sex chromosome aneuploidies, which commonly are not lethal but are responsible for disorders of sex development. Balanced chromosome rearrangements cause impaired fertility due to the mortality of embryos with an unbalanced chromosome complement, http://www.csun.edu/~cmalone/pdf360/Ch08-2%20number.pdf

The most likely explanation for the low rate of success may be failures in epigenetic reprogramming and imprinting - mechanisms to make sure that a zygote ends up with a 'clean slate' from two different parents, and that the 'interests' of each parent are balanced.

- Imprinted genes are mostly involved in fetal, placental and neonatal growth, with "paternally derived genes tending to foster large offspring, while maternally derived genes would regulate growth to safeguard the mother", according to the so-called "Genetic conflict" hypothesis.
- "Imprinting" is normally accomplished during spermatogenesis and oogenesis, processes that in humans take months and years, respectively, Figure (59). During SCNT (somatic cell nuclear transfer), reprogramming of the somatic donor nucleus must take place within minutes or, at most, hours between the time that nuclear transfer is completed and the onset of cleavage of the activated egg begins. It is known that the cytoplasm of an oocyte can reprogram the genome of a somatic cell to an embryonic state, but the precise mechanisms are not clear.
- Reprogramming of Epigenetic is a normal process in gametes and early embryo (and must also happen in SCN Tembryos) that re-sets the correct embryonic model of gene expression. As cells begin to differentiate, cell-type-specific patterns of DNA methylation and histone adaptations are acquired".

Figure (77): Normal and SCNT cloning

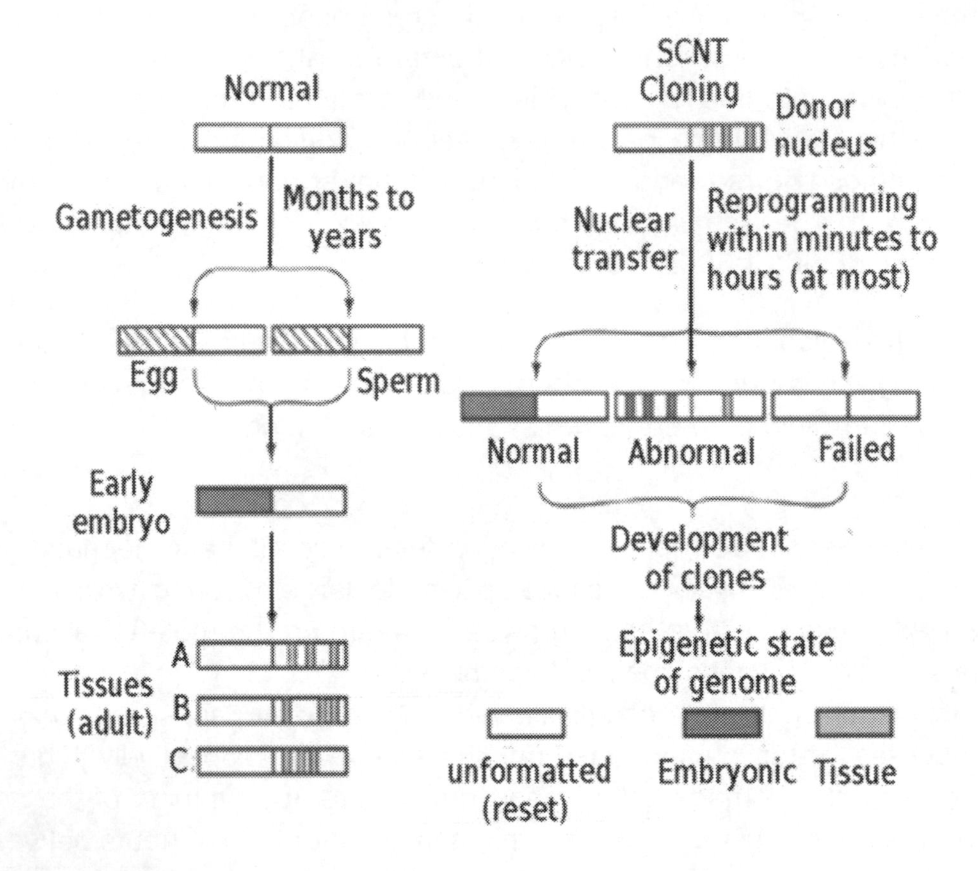


## 4.7.2 Ethical Issues of Cloning

In bioethics, the ethics of cloning refers to a variety of ethical positions regarding the practice and possibilities of cloning, especially human cloning. While many of these views are religious in origin, some of the questions raised by cloning are faced by seclar perspectives as well. The American Association for the Advancement of Science (AAAS) and other scientific organizations have made public statements suggesting that human reproductive cloning be banned and excluded until safety issues are resolved. Serious ethical concerns have been brought up by the future possibility of harvesting organs from clones.

Advocates support development of therapeutic cloning in order to generate tissues and whole organs to treat patients who otherwise cannot obtain transplants, to avoid the need for immune suppressive drugs, and to stave off the effects of aging. Advocates for reproductive cloning believe that parents who cannot otherwise procreate should have access to the technology.

Opponents of cloning have concerns that technology is not yet developed enough to be safe, that it could be prone to abuse (leading to the generation of humans from whom organs and tissues would be harvested), and have concerns about how cloned individuals could integrate with families and with society at large.

Religious groups are divided, with some opposing the technology as usurping God's place and, to the extent embryos are used, destroying a human life; others support therapeutic cloning's potential life-saving benefits.

There are also ethical objections. Article 11 of UNISCO's Universal Declaration on the Human Genome and Human Rights asserts that the reproductive cloning of human beings is contrary to human dignity, that a potential life represented by the embryo is damaged when embryonic cells are used, and there is a significant likelihood that cloned individuals would be biologically damaged, due to the inherent unreliability of cloning technology.

### 4.7.3 Religious Views

Many conservative Christian groups have disagreed with human cloning and the cloning of human embryos, since they believe that life begins at the moment of conception Other Christian denominations such as the United Church of Christ do not believe a fertilized egg constitutes a living being, but still they oppose the cloning of embryonic cells. The Roman Catholic Church, under the papacy of Benedict XVI, condemned the practice of human cloning, in the magisterial instruction Dignities Personae, stating that it represents a "grave offense to the dignity of that person as well as to the fundamental equality of all people. The World Council of Churches, representing nearly 400 Christian denominations worldwide, opposed cloning of both human embryos and whole humans in February 2006. The United Methodist Church opposed research and reproductive cloning in May 2000 and again in May 2004.

The prominent Qatari Islamic scholar, Yusuf Al Qaradawii judjes that cloning specific parts of the human body for medical purposes is not prohibited in Islam, but to clone the whole human body would not be permitted under any circumstances but on the issue of animal ethics he takes a more lenient position.

The late Grand Ayatollah of Lebanon, Mohammad Hussein Fadlallah did not see cloning as illegal. He also stressed that Islam supports the pursuit of the sciences including medicine. The Ayatollah did however warn against cloning the entire human being for the purpose of harvesting his or her organs.

Sunni Muslims consider human cloning to be banned by Islam. The Islam Fiqh Academy, in its Tenth Conference proceedings, which was convened in Jeddah, Saudi Arabia in the period from June 28, 1997 to July 3, 1997, issued a Fatwa stating that human cloning is haram (forbidden and sinful).

Judaism does not equate life with conception and, though some question the wisdom of cloning, Orthodox rabbis generally find no firm reason in Jewish law and ethics to object to cloning. Liberal Jewish thinkers have warned against cloning, among other genetic engineering efforts, though some prize the potential medical advantages.

Those who plan to clone humans have suggested that they will take additional precautionary steps beyond those presently undertaken in animal cloning. The steps include preimplantation testing to detect chromosome defects and errors in imprinting (methylation) at one or more DNA sites, and postimplantation testing of the imprinting (methylation) status at up to 30 DNA sites. All participants would sign an informed-consent form that would outline the dangers to both the mother and the child and the low probability of success. Those who have publicly stated their intention to undertake human reproductive cloning are thus far using private funding in a nonuniversity setting, and in some cases they are operating or planning to operate outside the United States.

### 4.7.4 Cloning extinct and endangered species

Around 100 species become disappeared a day. Despite increasing interest in using cloning to rescue endangered species, successful interspecies nuclear transfer has not been previously described, and only a few reports of in vitro embryo formation exist. Here we show that interspecies nuclear transfer can be implemented to clone those in danger of extinction species with normal karyotypic and phenotypic development through implantation and the late stages of fetal growth. Somatic cells from a gaur bull (Bos gaurus), a large wild ox on the verge of extinction, (Species Survival Plan < 100 animals) were electrofused with enucleated oocytes from domestic cows. Twelve to fifteen percent of the reconstructed oocytes developed to the blastocyst stage, and 18% of these embryos developed to the fetal stage when transferred to surrogate mothers. Three of the fetuses were electively removed at days 50 of gestation, and two continued gestation longer than 180 (ongoing) and 200 days, respectively. Microsatellite marker and cytogenetic analyses confirmed that the nuclear genome of the cloned animals was gaurus in origin. The gaur nuclei were shown to direct normal fetal development, with differentiation into complex tissue and organs, even though the mitochondrial DNA (mtDNA) within all the tissue types evaluated was derived exclusively from the recipient bovine oocytes. These results suggest that somatic cell cloning methods could be used to restore endangered, or even extinct, species and populations.

# CHAPTER 5

# STEM CELLS

Embryonic stem cells come from embryos that are three to five days old. At this stage, an embryo is called a blastocyst and has about 150 cells. In general, Stem cells are mother cells that have the potential to become any type of cell in the body. One of the main characteristics of stem cells is their ability to self-renew or multiply while maintaining the potential to develop into other types of cells. Stem cells can become cells of the blood, heart, bones, skin, muscles, brain etc. There are different sources of stem cells but all types of stem cells have the same capacity to develop into multiple types of cells, Figure (78),

Stem cells are differentiated from other cell types by two important characteristics. First, they are unspecialized cells capable of renewing themselves through cell division, sometimes after long periods of inactivity. Second, under specific physiologic or experimental conditions, they can be induced to become tissue- or organ-specific cells with special functions. In some organs, such as the gut and bone marrow, stem cells regularly divide to repair and replace worn out or damaged tissues. In other organs, however, such as the pancreas and the heart, stem cells only divide under special conditions.

Early in life, stem cells have the extraordinary potential to develop into any type of cell in the human body. They start in the embryo as unprogrammed cells, then become specialized to create bone, muscle, skin, the heart, the brain, and over 250 other types of specialized cells. These are called pluripotent stem cells.

Researchers have found that stem cells can be used to treat disease and injury. They stimulate the body to repair itself.

Researchers are finding new ways to use stem cells to rebuild tissue in many parts of the body where it has been damaged, such as the eye, the pancreas and the brain. For example, bone marrow transplants have been taking place for more than 40 years. These procedures rely on transplanting stem cells derived from bone marrow and have dramatically altered the treatment of blood disorders and certain cancers such as leukemia. In the past 20 years, significant new discoveries have emerged — breakthroughs that the original discoverers of stem cells never dreamed about. Researchers are finding new ways to use stem cells to rebuild tissue in many parts of the body where it has been damaged, such as the eye, the pancreas and the brain. Some revolutionary treatments for blindness, MS, stroke and spinal cord injury are already in early stage clinical trials.

Figure (78): Stem cells develop into multiple types of cells

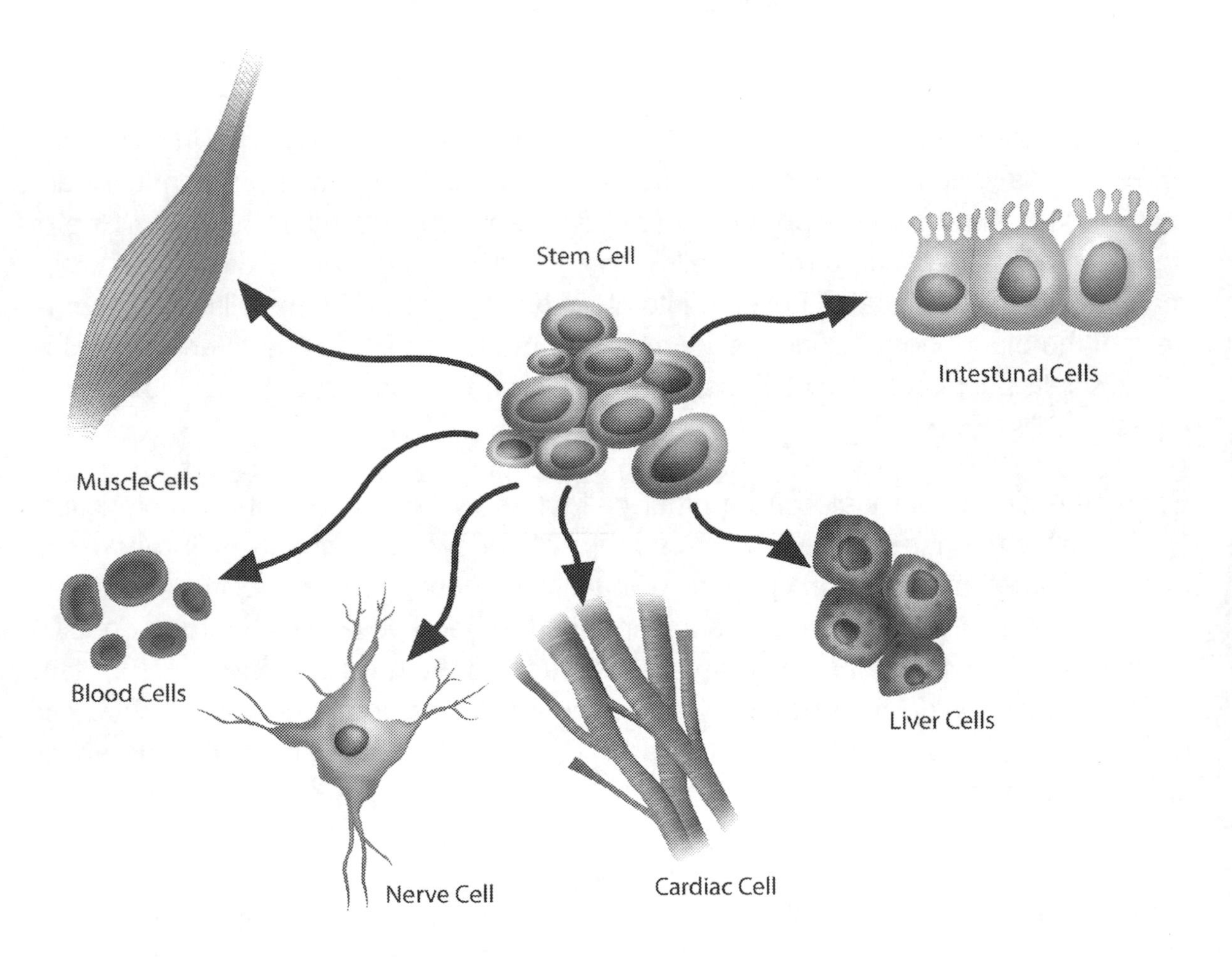


## 5.1 Unique Properties of Stem Cells

Stem cells are different from other kinds of cells in the body. All stem cells—regardless of their source—have three general properties: they are capable of dividing and renewing themselves for long periods; they are unspecialized; and they can change to specialized cell types,

Stem cells are unspecialized. specialised cells, such as blood and muscle cells, are unable to divide. A stem cell does not have any specialized physiological properties.

Stem cells can divide and produce identical copies of themselves over and over again. This process is called self - renewal and continues throughout the life of the organism. Self-renewal is the defining property of stem cells. Stem cells can also divide and produce more specialized cell types. This process is called differentiation. Stem cells from different tissues, and from different stages of development, vary in the number and types of cells that they can produce.

Stem cells have a special ability to divide and renew themselves for extended periods of time:

1. Why can embryonic stem cells proliferate for a period of more than a year or more in the laboratory without differentiating, but most adult stem cells cannot; and
2. Scientists try to know the factors in living organisms that normally regulate stem cell proliferation and self-renewal.

Discovering the answers to these questions may make it possible to understand how cell proliferation is regulated during normal embryonic development or during the abnormal cell division that leads to diseases such as cancer. Such information would also enable scientists to grow embryonic and non-embryonic stem cells more efficiently in vitro.

## 5.2 Embryonic stem cells

Embryonic stem cells are stem cells taken from the undifferentiated inner mass cells of a human embryo. Embryonic stem cells are pluripotent, meaning they are able to grow (i.e. differentiate) into all derivatives of the three primary germ cells: ectoderm, endoderm and mesoderm.

Embryonic stem cells, as their name suggests, are derived from embryos. Most embryonic stem cells are derived from embryos that develop from eggs that have been fertilized by eggs in vitro, and then donated for research purposes with informed consent of the donors. They are not derived from eggs fertilized in a woman's body.

As long as the embryonic stem cells in culture are grown under appropriate conditions, they can remain undifferentiated (unspecialized). But if cells are allowed to clump together to form embryoid bodies, they begin to differentiate spontaneously. They can form derm cells, muscle cells, nerve cells, and many other cell types, Figure (79).

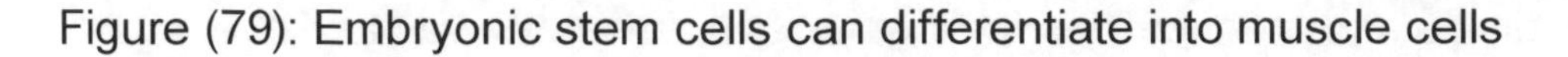

Figure (79): Embryonic stem cells can differentiate into muscle cells

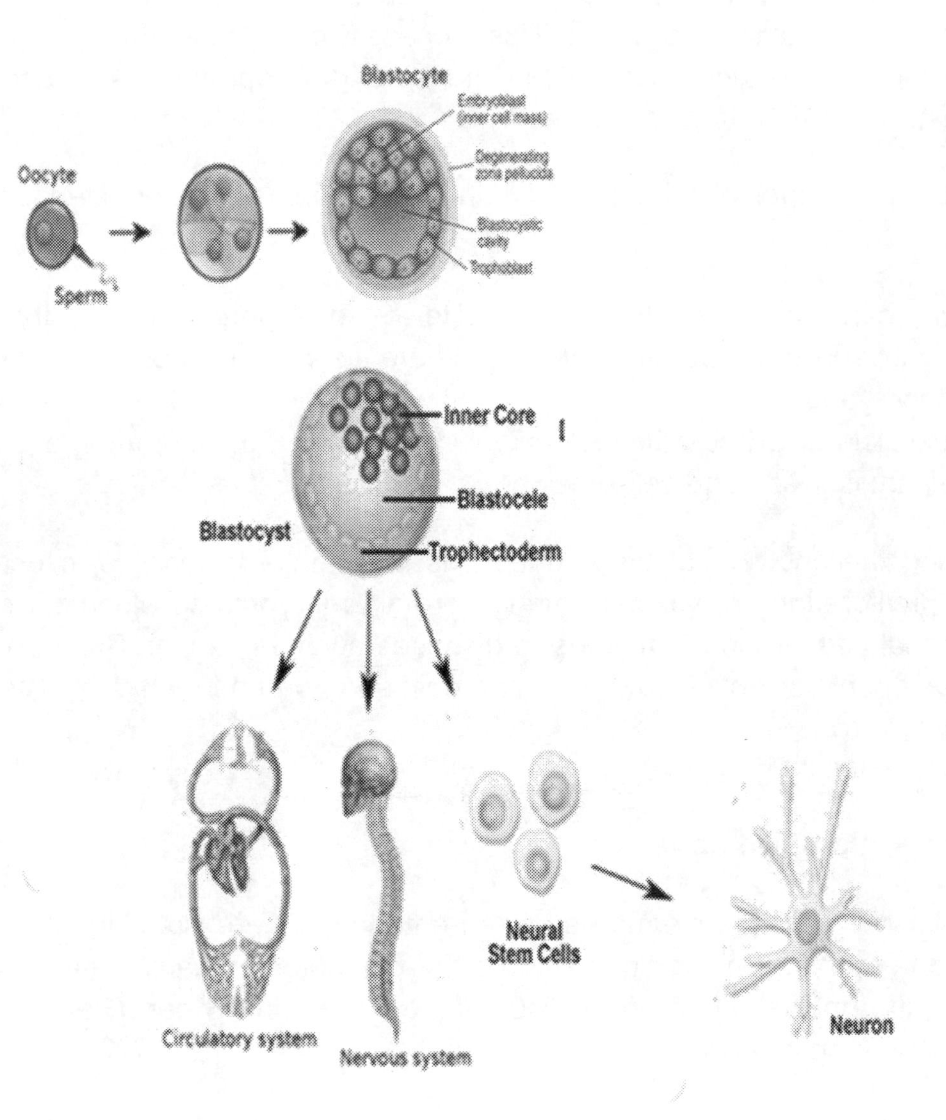


## 5.3 Adult Stem Cells

Adult stem cells are often present in only minute quantities and can therefore be difficult to isolate and purify. There is also evidence that they may not have the same capacity to multiply as embryonic stem cells. They don't have the development capability as those in embryonic stem cells, Figure (80).

Adult stem cells may contain more DNA abnormalities – caused by environment, sun, toxins, and errors in making more DNA copies during the course of a life time. These potential weakness might limit the usefulness of adult stem cells.

Totipotent cells: These are the most versatile of the stem cell types. The fertilized egg is totipotent, meaning it has the potential to give rise to any and all human cells, such as heart, brain, liver, or blood cells.

Pluripotent cells: Pluripotent, embryonic stem cells originate as inner mass cells within a blastocyst. These stem cells can become any tissue in the body, excluding a placenta. Only the morula's cells are totipotent, able to become all tissues and a placenta.. Embryonic stem (ES) cells are pluripotent cells. Multipotent cells: These cells, can only give rise to cell types within their lineage. Most adult stem cells are multipotent cells.

Figure (80): Embryonic stem cells and somatic cells develop into endoderm, mesoderm and ectoderm

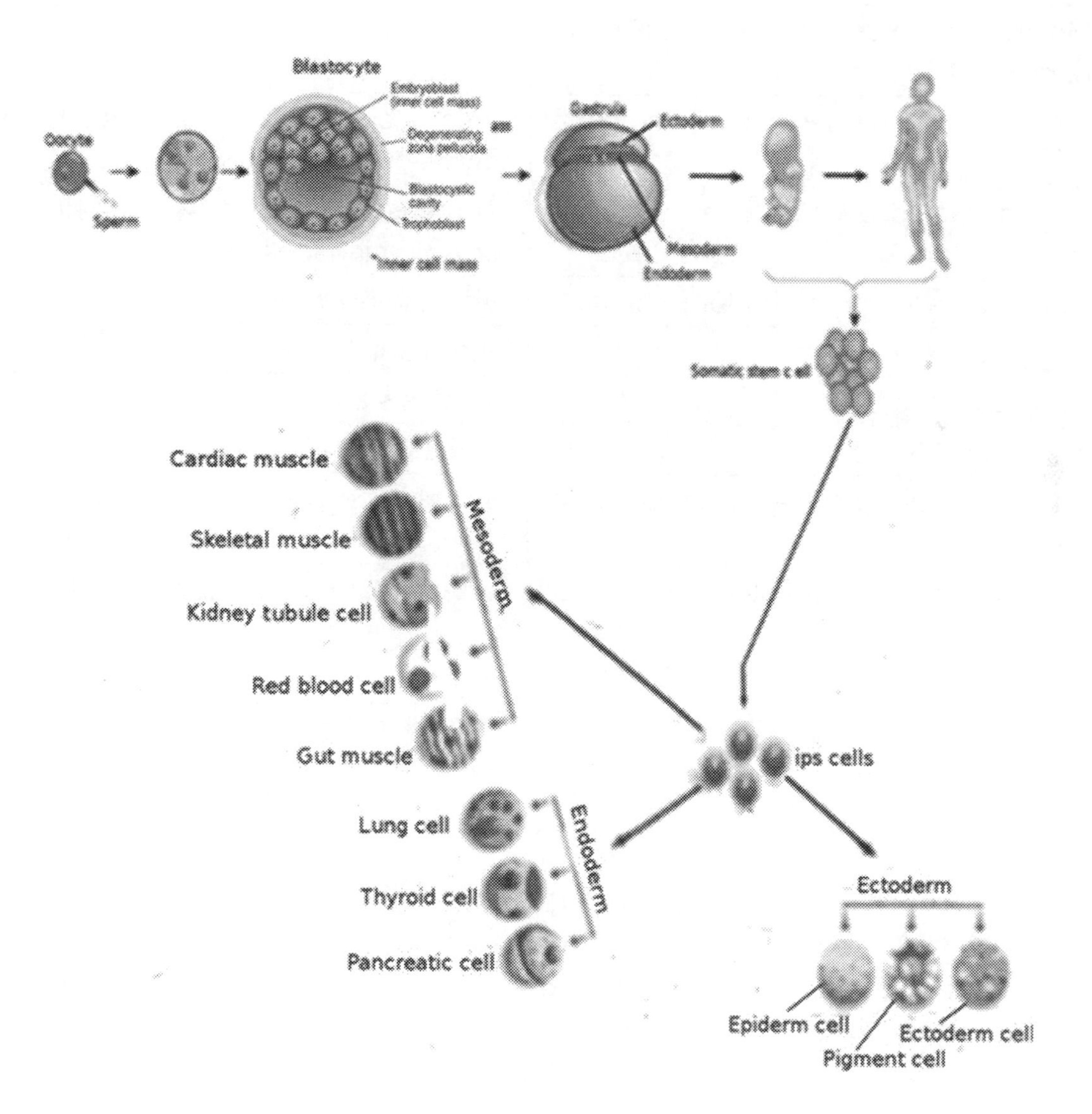

When a sperm meets an egg, the resulting zygote is totipotent. The inner cell mass which is the source of "embryonic stem" cells are pluripotent.

Embryonic stem cells (ESC) can differentiate into any cell type (totinpotent/pluripotent) while adult Stem cells (ASC) have already committed to a particular fate (multipotent)..

All stem cells are undifferentiated, meaning that they have not become a "type" of cell yet.

Table (1) shows the difference between human adult stem cells and human embryonic stem cells

| Human adult stem cells | Human embryonic stem cells |
|---|---|
| Mostly Multipotent with MSGs acting as pluripotent | Totinpotent/pluripotent |
| Stem cells are hard to access and purify | Once isolated, the cells show high degree of proliferation |
| Chromosomes tend to shorten with aging | Chromosome length is maintained across serial passage |
| Telomerase length is shorter with aging | Telomerase levels high |
| No teratoma risk | Significant Teratoma risk |
| Apoptosis mayt be early | Apoptosis is late |
| No ethical Issues | Serious ethical issues |
| Patient – specific hence less chances of immune rejection | High chance of immune rejection |
| Small amount of pluripotent cells. which can give riser to most types of tissues, and definitely not a whole organism | Large amount of pluripotent cells |
| Unipotent can regenerate but can only differentiate into their associated cell type (e.g. liver,, kidney, heart, blood cells, etc. | Totipotent cells in zygotes and early morula retain ability |
| Induced pluripotent stem cell (IPSCs) is adult cells that have been genetically programmed to any embryonic stem cells. | |
| Human adult stem cells can be found in bone marrow, umbilical cord blood, mammary glands, intestine, mesenchymal tissues, neural cells, testicular cells and endothelial cells | Zygote and embryo |

## 5.4 Induced Pluripotent Stem Cells

As we pointed out earlier, induced pluripotent stem cells (iPSCs) are pluripotent stem cells generated from adult cells by reprogramming. iPSCs have the same properties as embryonic stem cells, and therefore self-renew and can differentiate into all cell types of the body except for cells in extra-embryonic tissues such as the placenta.

Induced pluripotent stem cells are a type of pluropotent stem cell that can be generated directly from adult cells. The iPSC technology was pioneered by Shinya Yamanaka's lab in Kyoto who introduced the iPS in 2006. He was awarded the 2012 Nobel Prize along with Sir John Gardon "for the discovery that mature cells can be reprogrammed to become pluripotent."

Pluripotent stem cells hold great promise in the field of regenerative medicine. Because they can propagate indefinitely, as well as give rise to every other cell type in the body (such as neurons, heart, pancreatic, and liver cells), they represent a single source of cells that could be used to replace those lost to damage or disease, Figure (81).

Although these cells meet the defining criteria for pluripotent stem cells, it is not known if iPSCs and embryonic stem cells differ in clinically significant ways. Mouse iPSCs were first reported in 2006, and human iPSCs were first reported in late 2007. iPSC derivation is typically a slow and inefficient process, taking 1–2 weeks for mouse cells and 3–4 weeks for human cells, with efficiencies around 0.01%–0.1%. However, considerable advances have been made in improving the efficiency and the time it takes to obtain iPSCs. Upon introduction of reprogramming factors, cells begin to form colonies that resemble pluripotent stem cells, which can be isolated based on their morphology, conditions that select for their growth, or through expression of surface markers or reporter genes.

Viruses are currently used to introduce the reprogramming factors into adult cells, and this process must be carefully controlled and tested before the technique can lead to useful treatment for humans. In animal studies, the virus used to introduce the stem cell factors sometimes causes cancers. Researchers are currently investigating non-viral delivery strategies. In any case, this breakthrough finding has created a powerful new way to "de-differentiate" cells whose developmental fates had been previously assumed to be determined. In addition, tissues derived from iPSCs will be a nearly identical match to the cell donor and thus probably avoid rejection by the immune system. The iPSC strategy creates pluripotent stem cells that, together with studies of other types of pluripotent stem cells, will help researchers learn how to reprogram cells to repair damaged tissues in the human body.

Figure (81): Pluripotent stem cells

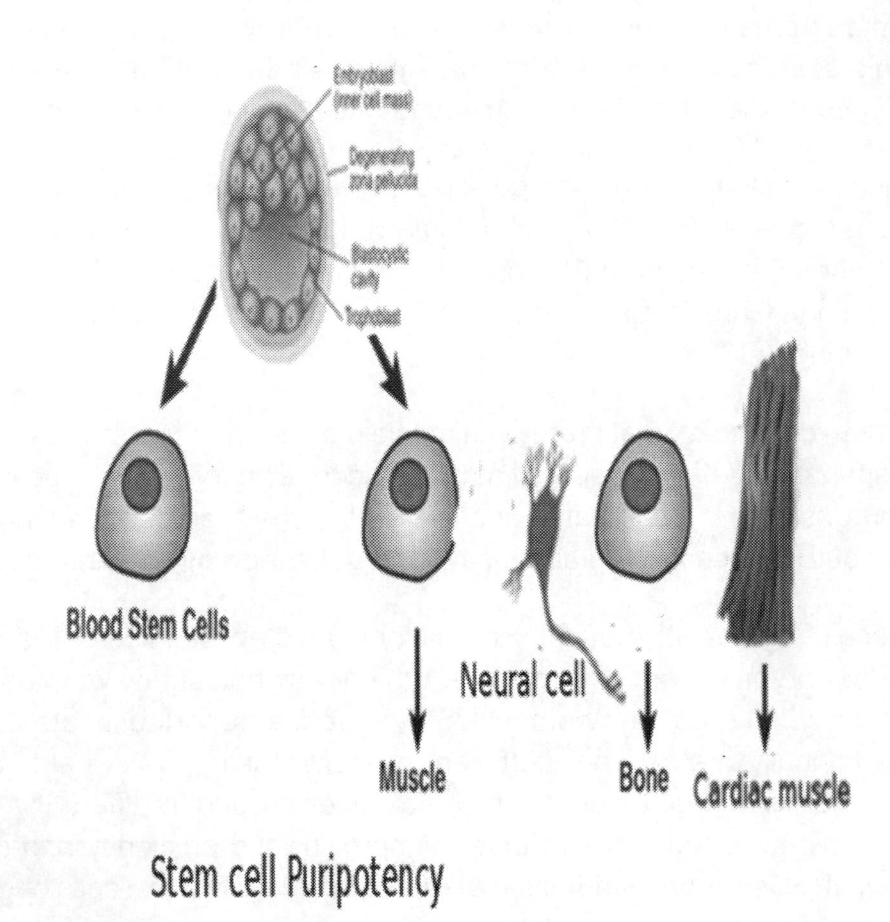


## 5.5 Current Arguments and Counterarguments Regarding Human Reproductive Cloning

This dissertation is about the 'stem cell debate' that is, the debate over the ethics of the generation and use of all types of stem cells, in particular human stem cells. The stem cell debate has an unprecedented importance. Provided below is a summary of some of the current arguments and counterarguments regarding human reproductive cloning, put by the National Bioethics of the Nationa Academy (USA) Advisory Commission. The panel's analysis of each is based on the scientific and medical literature and on presentations at its workshop.

Argument 1: Animal-safety data do not apply, because humans are very different from the animals under study. In particular, a recent study on monkeys. Testing on monkeys at 500 times the dose given to the volunteers totally failed to predict the dangerous side effects.

Counterargument: Placental function, development, and genetic regulation are the same in humans and animal models, such as mice, so similar SCNT-related defects would be predictable. Several studies have emphasized that humans and other organisms have the same basic pathways for governing early embryonic and fetal development. Furthermore, extensive defects in all five of the mammalian species that have been reproductively cloned thus far suggest that the defects would affect basic biological functions in humans.

Even if one less gene is imprinted in humans as compared to mice, humans are known to have many imprinted genes (possibly as many as 100), and any number of these are likely to cause troubles in reproductively cloned humans.

Argument 2: Frequent malfunctions are seen in normal human reproduction; cloning would be no different.

Counterargument: Errors in normal human reproduction occur primarily early in pregnancy; many of the women in question are never aware that they are pregnant. In contrast, many of the defects in reproductively cloned animals appear late in pregnancy or after birth.

Argument 3: improper culture media for the initial cells cause most cloning-related problems. Culture media for human pre-implantation embryos in assisted reproductive. Culture media for human assisted reproductive technologies have been better optimized. Synchronization between the implanted embryo and the recipient uterus has also been better in human than in animal assisted reproductive technology procedures.

Counterargument: Culture effects appear to account for only some of the defects observed. Many defects in various organ systems are abnormal to reproductive cloning. Expertise in existing human assisted reproductive technologies is not relevant to these problems, because the defects appear to arise from biological rather than purely technical causes.

Argument 4: Those who have cloned animals stress the failures, but there are also many successes in animal reproductive cloning. However, to date, the story of cloning endangered animals is one of a few high-profile successes and many, many failures.

Counterargument: The statement is factual, but does not necessarily apply to human reproductive cloning. In humans, the likelihood and benefit of success must be weighed against the probability, severity, and lifelong consequences of failure. Failures are all but certain in any human reproductive cloning attempt at this time, based on the experience with animals, and in humans, the consequences could be far more dreadful. The likelihood and benefit of possible success must be weighed against the high probability and severe consequences of failure.

Argument 5: Preimplantation Genetic Diagnosis (PGD) is reviewed and novel fields where it may be applied are investigated. Existing preimplantation and postimplantation genetic tests could be used to detect abnormalities, allowing selection of embryos to be implanted and therapeutic abortion in case of any problems. In contrast, there has been no genetic testing and weeding out of animal reproductive clones.

Preimplantation genetic diagnosis is the process of removing a cell from an in vitro fertilization embryo for genetic testing before transferring the embryo to the uterus. In preimplantation testing, two cells could be removed from an eight-cell morula. One cell could be tested for correctness of the chromosome complement and the other for imprinting errors at one or more DNA sites. It has been claimed that such imprinting tests have been performed with DNA from cells after somatic cell nuclear transfer (SCNT), although no data have been presented. Postimplantation testing could include testing for chromosomal errors, the checking of imprinting status at up to 30 sites, and the measurement of production levels from many genes with DNA chips or reverse-transcription polymerase chain reaction.

Counterargument: Many errors would not be demonstrable until late in pregnancy or after birth, when therapeutic abortion would not be an option. Many of the relevant genetic tests have not yet been developed; existing genetic tests appropriate for single-gene inherited disorders or gross chromosomal rearrangements are not enough because they are not relevant to the major sources of errors expected in human cloning. Ultrasonographic evaluation cannot detect the small-scale defects in tissues, such as lung, that have had devastating consequences in newborn animal clones, and there is not enogh evidence regarding the possible impact of imprinting errors on brain development in humans.

Argument 6: Obtaining informed consent for medical treatment, for participation in medical research, and for participation in teaching exercises involving students and residents is an ethical requirement that is partially reflected in legal doctrines and requirements. Voluntary informed consent allows potential participants to make their own decisions and elect to take the risks if they so choose.

Counterargument: The current regulatory system recognizes that when information is lacking it can be difficult or impossible to inform subjects fully. That is the case with respect to human reproductive cloning because the extent of the risks is unknown, and the greatest risk ofdeformity, morbidity, and mortality is borne by the cloned fetus/child, who cannot give informed consent. In addition, there are dangers borne by the woman donating the eggs and the gestational mother.

When subject matter cannot be fully informed, and when a procedure is clearly risky, there is a role for both regulatory agencies and professionals to limit the options available to a subject if the evidence supports such a limitation. Societal concerns can also be taken into account.

Criminalizing the implantation step should be adequate to avoid such proscribed activity. Moreover, because all nuclear transplantation experiments will require the participation of human subjects (the donor of the eggs and the donor of the somatic cell nuclei, who may be the same person or different persons), all this work would necessarily be regulated and controlled by the procedures and rules concerning human-subjects research—subjecting it to close scrutiny.

Stem cells originated directly from an adult's own tissues are an alternative to nuclear transplantation-derived embryonic stem cells as a source of cells for therapies. Two types of adult stem cells—bone marrow and skin stem cells—currently provide the only two stem cell therapies. But, as noted in the above mentioned report, many questions remain before the potential of other adult stem cells can be perfectly assessed. Few studies on adult stem cells have adequately defined the stem cell by starting from a single isolated cell or defined the necessary cellular environment for accurate differentiation or the factors controlling the efficiency with which the cells repopulate an organ. There is a need to show that the cells derived from introduced adult stem cells are contributing directly to tissue function and to improve the ability to maintain adult stem cells in culture without having the cells distinguished. Finally, most of the studies that have garnered so much attention have used mouse rather than human adult stem cells.

The preceding report also notes that unlike adult stem cells, it is well confirmed that embryonic stem cells can form multiple tissue types and be maintained in culture for long periods of time. However, embryonic stem cells are not without their own potential problems as a source of cells for transplantation. The growth of human embryonic stem cells in vitro now needs a "feeder" layer of mouse cells that may contain viruses, and when allowed to differentiate the embryonic stem cells can form a mixture of cell types at once. Human embryonic stem cells can form benign tumors when introduced into mice, although this potential seems to vanish if the cells are allowed to differentiate before introduction into a recipient.

In addition to possible uses in therapeutic transplantation, embryonic stem cells and cell lines derived by nuclear transplantation could be significant tools for both fundamental and applied medical and biological research. This research would begin with the transfer of genetically defined donor nuclei from normal and diseased tissues. The resulting cell lines could be used to study how inherited and acquired alterations of genetic components might contribute to disease processes. The properties of the cell lines could be studied directly, or the embryonic stem cells could be examined as they differentiate into other cell types. For instance, the way in which cells derived by nuclear transplantation from an heimer's disease patient acted while differentiating into brain cells, compared with those derived from a normal patient, might bring in new clues about Alzheimer's disease. Such cell lines could also be used to ensure that research covers a more genetically diverse human population than that represented in the blastocysts stored in IVF clinics, promoting studies of the causes and consequences of genetic diseases by letting allowing researchers to study how embryonic stem cells with different genetic endowments differ in the way that they form cell types and tissues. To conclude, studies of genetic reprogramming and genetic imprinting will be substantially enhanced through the use of stem cells derived by nuclear transplantation, compared with studies with stem cells derived from other sources,

## 5.6 Summary

Most of the relevant data on reproductive cloning are derived from animal studies. The data reveal high rates of abnormalities in the cloned animals of multiple mammalian species and lead the panel to bring about that reproductive cloning of humans is not now safe and not reliable. Our present opposition to human reproductive cloning is based on science and medicine, irrespective of broader considerations. However, that a broad ethical debate must be encouraged, so that the public can be prepared to make choice if human reproductive cloning is some day considered medically safe for mothers and offspring.

In their discussion, the outcome should inevitably include a comparison of the methods used for reproductive cloning and for nuclear transplantation to produce stem cells.

CHAPTER 6

# GENE THERAPY AND CLONING THERAPY

## 6.1 Gene Therapy

Gene therapy is an experimental technique that uses genes to treat or prevent disease. Gene therapy is when DNA is introduced into a patient to treat a genetic disease. The new DNA usually contains a functioning gene to correct the effects of a disease-causing mutation.

- Gene therapy uses sections of DNA (usually genes) to treat or prevent disease.
- The DNA is carefully selected to correct the effect of a mutated gene that is causing disease.
- The origins of gene therapy can be traced back to the first live attenuated vaccines in the 1950s. The technique was first developed in 1972 but has, so far, had limited success in treating human diseases.
- Gene therapy may be a promising treatment option for some genetic diseases, including muscular dystrophy and cystic fibrosis.

Researchers are testing several approaches to gene therapy, including:

- Replacing a mutated gene that causes disease with a healthy copy of the gene.
- Inactivating, or "knocking out," a mutated gene that is functioning improperly.
- Introducing a new gene into the body to help fight a disease.

In somatic cell gene therapy (SCGT), the therapeutic genes are transferred into any cell other than a gamete, germ cell, gametocyte or undifferentiated stem cell.

In germline gene therapy (GGT), germ cells (sperm or eggs) are modified by the introduction of functional genes into their genomes. Modifying a germ cell causes all the organism's cells to contain the modified gene.

There are two different types of gene therapy depending on which types of cells are treated:

1. Somatic gene therapy: transfer of a section of DNA to any cell of the body that doesn't produce sperm or eggs. Effects of gene therapy will not be passed onto the patient's children.
2. Germline gene therapy: transfer of a section of DNA to cells that produce eggs or sperm. Effects of gene therapy will be passed onto the patient's children and subsequent generations.

Gene therapy is an experimental technique that uses genes to treat or prevent disease. In the future, this technique may allow doctors to treat a disorder by inserting a gene into a patient's cells instead of using drugs or surgery. Researchers are testing a number of approaches to gene therapy, including:

Replacing a mutated gene that causes disease with a healthy copy of the gene Inactivating, or "knocking out," a mutated gene that is functioning improperly, or Introducing a new gene into the body to help fight a disease.

Although gene therapy is a promising treatment option for a number of diseases the technique remains risky and is still under study to make sure that it will be safe and effective. Gene therapy is currently only being tested for the treatment of diseases that have no other cures.

## 6.2 Somatic Gene Therapy

Somatic cell gene therapy changes genes in just one person. The targeted cells are the only ones affected; the changes are not passed on to that person's offspring. Germ line gene therapy makes changes in the sperm or egg of an individual.

The cells containing the vector are then returned to the patient. If the treatment is successful, the new gene delivered by the vector will make a functioning protein. Figure (82) shows Somatic gene therapy.

Figure (82): Somatic gene therapy

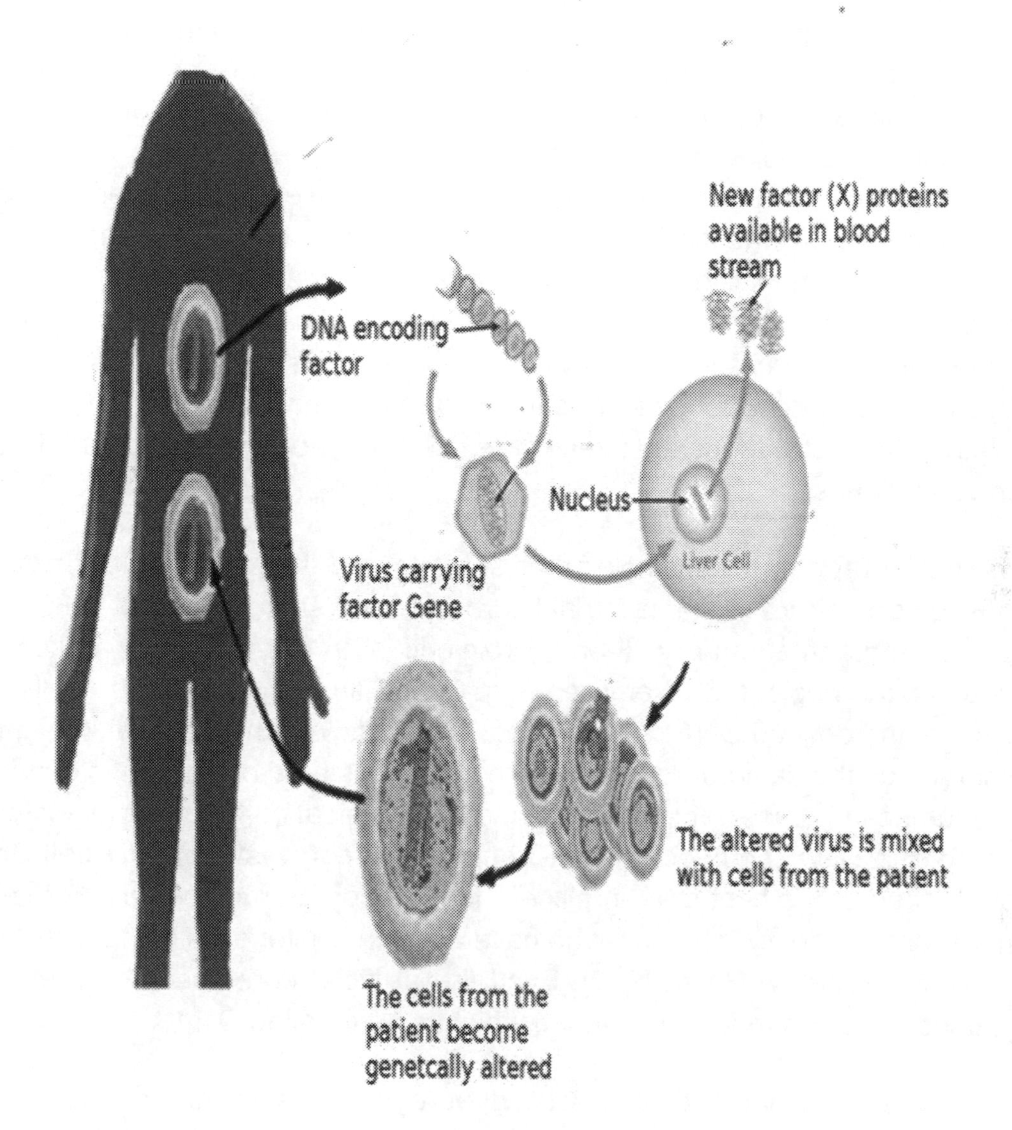


## 6.3 Germline Gene Therapy

Germline gene therapy involves the modification of germ cells (gametes) that will pass the change on to the next generation. With germline therapy genes could be corrected in the egg or the sperm that is being used to conceive. The child that results would be spared certain genetic problems that might otherwise have occurred. Because every cell descends from the fertilized egg, every cell in the offspring would possess the transplanted gene. This would be a far more effective way of transferring genes than the ones presently used in somatic cell therapies, where genes into the cells of children or adults usually enter only a small portion of the person's cells and eventually stop functioning. Germ-line gene therapy inserts genes into reproductive cells or possibly into embryos to correct genetic defects that could be passed on to future generations.

Policy, advancement of research, and decision-making regarding stem cells varies between states and between countries, due to ethical considerations, economic concerns, cultural concerns, religious beliefs, and personal values of their citizens. The benefits of germ line therapy seem limitless. Every year many genes that lead to medical disorders are discovered. In 1997 genes were discovered for epilepsy, mental retardation, Parkinson's disease, breast cancer, heroin addiction, glaucoma and obsessive-compulsive disorder to name just a few (Glausiusz 1997). Germ line therapy may be able to help cure these diseases in the present population and eradicate it from future generations. Adult body cells are much more abundant than embryonic cells, and no fetuses are sacrificed to make new tissue. Only the original colonies of stem cells, which are already in existence, need to be used for future organ cloning techniques.

Germ line therapy is also simpler that somatic cell therapy. It is already a well-established procedure in animals that has been used since the early 1980s Cloning is more accurate in an embryo than microinjection. In microinjection, naked DNA is injected into the nucleus of a fertilized egg. Only a small portion of the cells actually incorporate the desired DNA. With cloning, scientists will know before implantation if an embryo has the desired trait. The downside of this type of therapy is that it is very new in humans and has yet to be perfected. One ethical concern is the vectors used to transfer the DNA. The vector DNA would also incorporate into the cell DNA and become a part of the host DNA in all cells in the body. Some see this as resulting in "humans that are not fully human-who have alien characteristics of viruses and other species" (Natural Law Party 1998). Even the artificial chromosomes that are now being used are often made from yeast, which is again alien DNA,

Third-party reproduction or donor-assisted reproduction is any human reproduction in which DNA or gestation is provided by a third party or donor other than the two parents who will raise the resulting child. This goes beyond the traditional father-maother model, although the third party's involvement is limited to the reproductive process and does not extend into the raising of the child. Third-party reproduction is used by couples unable to reproduce in the traditional manner, including same sex and those where one or both partners is infertle.

One can distinguish several categories, some of which may be combined:

- Sperm donation. A donor provides sperm
- Egg donation. A donor provides ova

- Spindle transfer. A third party's mitochondrial DNA is transferred to the future mother's ovum. This is used to prevent mitochondrial disease.
- Embryo donation with embryos which were originally created for a genetic mother's assisted pregnancy. Once the genetic mother has completed her own treatment, she may donate unused embryos for use by a third party, or where embryos are specifically created for donation using donor eggs and donor sperm.
- Surrogacy. The embryo is gestated in a third party's uterus.
- Adoptation is usually considered separately from third-party reproduction.

The phenotype, the physical appearance, was the one of the natural mother and father but the chemical engine in his cells would run on a different mother mitochondria, Figure (83). In conclusion, Gene therapy defines as the insertion of normal or genetically altered genes into cells usually to replace defective genes especially in the treatment of genetic disorders.

Figure (83): Germline Gene Therapy

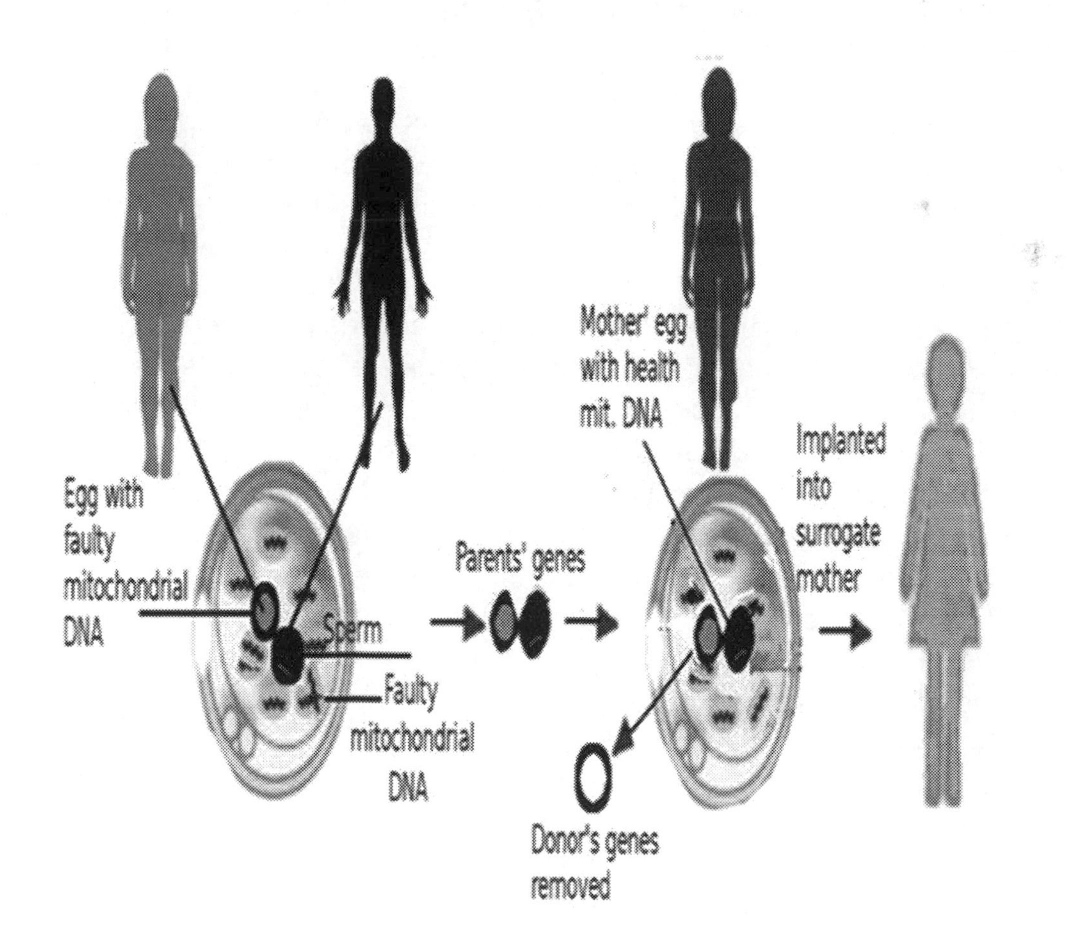

## 6.4 Cloning Therapy

Cloning therapy refers to the removal of a nucleus, which contains the genetic material, from virtually any cell of the body (a somatic cell) and its transfer by injection into an unfertilised egg from which the nucleus has also been removed. The newly reconstituted entity then starts dividing. After 4-5 days in culture, embryonic stem cells can then be removed and used to create many embryonic stem cells in culture. These embryonic stem cell 'lines' are genetically identical to the cell from which the DNA was originally removed, Figure (84).

Therapeutic Cloning does not involve making a clone of someone- you don't end up with a cloned person: that's reproductive cloning. Therapeutic cloning involves creating copies of cells to make someone better.

The first thing to say is that an embryo contains stem cells, and these are cells that can grow into any sort of cell in the body. This is very useful if you want to regrow dead or damaged cells, or grow entire organs. There are other ways to get stem cells, but some researchers think that this is the best way. Secondly, if you have an embryo with the same DNA as the patient, you are less likely to have problems with the patient's body rejecting the new cells.

Therapeutic cloning is still very new, although scientists have been able to develop the process in animals. Researchers believe that it could be used to treat a wide range of inherited disorders and conditions. These include Parkinson's, spinal cord injuries, MS, etc.

But one major problem with developing disease therapies with stem cells is the body's immune response system. When cells, including donated organs, tissues or blood, are transplanted or transfused, the recipient's body mounts a rejection response, attacking these cells as foreign. Advances in biotechnology necessitate both an understanding of scientific principles and ethical implications to be clinically applicable in medicine. In this regard, therapeutic cloning offers significant potential in regenerative medicine by circumventing immunorejection, and in the cure of genetic disorders when used in conjunction with gene therapy. Therapeutic cloning in the context of cell replacement therapy holds a huge potential for *de novo* organogenesis and the permanent treatment of Parkinson's disease, Duchenne muscular dystrophy, and diabetes mellitus as shown by *in vivo* studies. Scientific roadblocks impeding advancement in therapeutic cloning are tumorigenicity, epigenetic reprogramming, mitochondrial heteroplasmy, interspecies pathogen transfer, low oocyte availability. Therapeutic cloning is also often tied to ethical considerations concerning the source, destruction and moral status of IVF embryos based on the argument of potential. Legislative and funding issues are also addressed. Future considerations would include a distinction between therapeutic and reproductive cloning in legislative formulations.

It would also be possible to use therapeutic cloning to grow organs for transplant. It can take years of waiting on a donor list until you get an organ for transplant. If you could grow a cloned organ, it would take less time, and your body would not reject the organ.

The nation's top scientists from The National Academies of Science and National Institutes of Health, as well as numerous Nobel Laureates attest to the scientific value of this research. A February, 2002 report from the National Academies of Science concluded that while reproductive cloning (creating humans) is unsafe and should be banned, therapeutic cloning has sufficient scientific potential that it should be allowed to continue.

Figure (84): Human and therapeutic cloning.

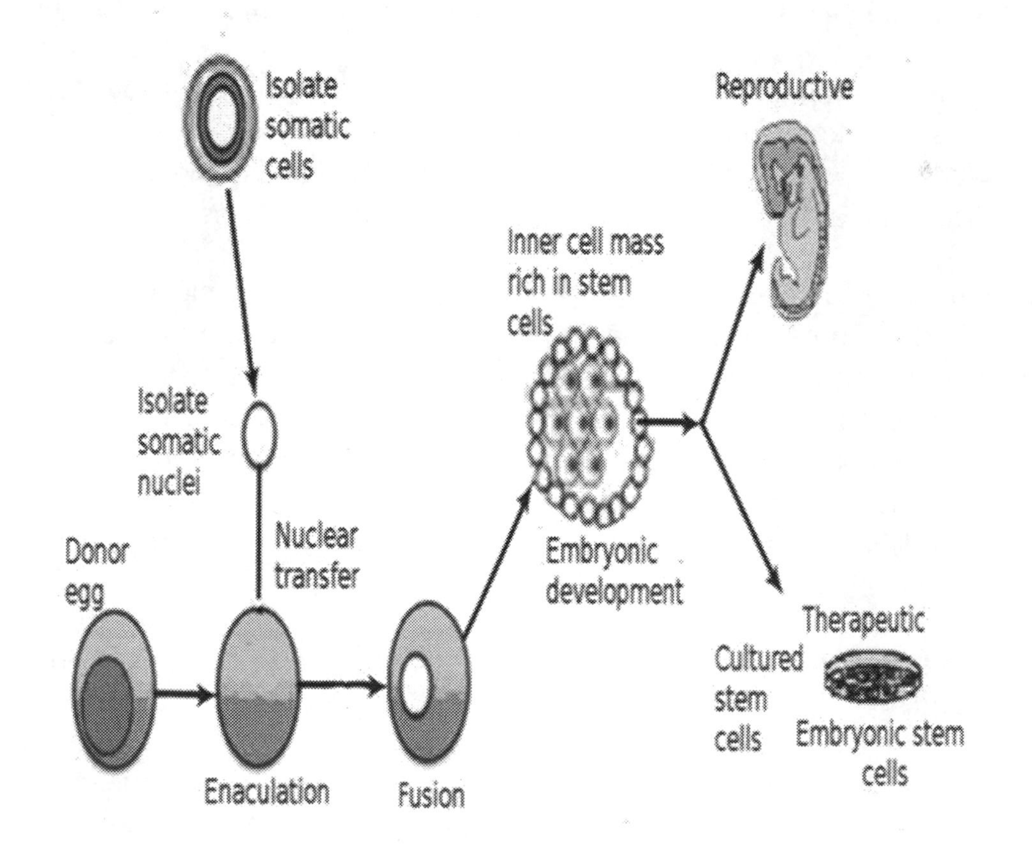

When people think of the word 'cloning', they are often hit with frightening images of duplicate human beings being created in somewhat of a mad scientist style experiment. Although several scientific roadblocks remain unsolved, the medical benefits that could be gained from treatments based on therapeutic cloning outweigh the ethical dilemma and calls for further improvements to be clinically applicable. In sum, therapeutic cloning features great potential as a histocompatible method for cell replacement therapy to restore motility following paralysis, counteract senescence, and repair damages done by stroke, myocardial infarction, liver cirrhosis, severe burns and osteoporosis to name a few.

## 6.5 Conclusion

Approval of the gene therapy paves the way for the development of treatments for more widespread illnesses such as thalassemia and sickle cell disease. Hundreds of inherited disorders such as cystic fibrosis, muscular dystrophy and many types of blindness are caused by faulty genes,. Gene therapy is an experimental technique that uses genes to treat or prevent disease. In the future, this technique may allow doctors to reverse aging by introducing a gene into a aged individuals to rejuvenate cells. Researchers are testing several approaches to gene therapy, including:

- Replacing a mutated gene that causes disease with a healthy copy of the gene.
- Replacing old genome with young ones.
- Inactivating, or detach a mutated gene that is malfunctioning.
- Introducing a new gene into the body to help fight a disease.

Genes, which are carried on chromosomes, are the basic physical and functional units of heredity. There are two types of genes: Germ line gene, i.e., sperm or eggs, are modified by the introduction of functional genes, which are ordinarily integrated into their genomes. Therefore, the change due to therapy would be heritable and would be passed on to later generations, and the Somatic gene therapy in which the therapeutic genes are transferred into the somatic cells of a patient. Any modifications and effects will be restricted to the individual patient only, and will not be inherited by the patient's offspring.

This new method, theoretically, should be highly effective in counteracting genetic disorders and hereditary diseases. However, many jurisdictions and ethical principles prohibit this for exercising in human beings, for a variety of technical and ethical reasons. Genes are specific sequences of bases (nucleotides) that program instructions on how to synthesize enzymes and proteins. Although genes get a lot of attention, it's the proteins that perform most life functions and even make up the majority of cellular structures. When genes are altered so that the encoded enzymes and proteins are unable to perform their normal functions, diseases and aging are the consequences.

Gene therapy is being thoroughly researched as a cure for several genetic diseases. Out of all the genetic disorders, gene therapy for both sickle cell and hemophilia diseases has the most favorable characteristics for this potential cure. Gene therapy works in hemophilia by using DNA as the drug and viruses as the deliverer. For example, a virus containing the gene that produces Factor VIII or Factor IV (in case of Hemophilia B) is injected into a large group of cells in the patient. Currently, researchers are attempting to engineer cells, usually from bone marrow, to enhance the abilities of immune cells to fight off cancer and resist infection by HIV. This approach may lead to an effective treatment for nonhereditary diseases. The hope of gene therapy is to have the cell produce more of the cured cells and spread throughout the rest of the body. If successful, the patient would never need factor replacement therapy again and would be cured of diseases.

Los Angeles times, October 30, 2009, quoted "Pennsylvania researchers using gene therapy have made significant improvements in vision in 12 patients with a rare inherited visual defect, a finding that suggests it may be possible to produce similar improvements in a much larger number of patients with retinitis pigmentosa and macular degeneration"

Age-related deterioration in critical brain networks may be restored by gene therapy, according to a study in monkeys presented at the American Academy of Neurology's 52$^{nd}$ Annual Meeting in San Diego, CA, April 29 -- May 6, 2000. This finding lends support to treat Alzheimer's disease using a similar gene therapy approach, say the study's authors. Researchers from the University of California in San Diego found that normal aging in monkeys causes a 28 percent decline in the density of certain brain networks originating from nerve cells called neurons deep in the brain. Book of Biochemistry of Aging by this Author.

The scientists discovered that they were able to restore these connections by transplanting brain cells genetically programmed to release a protein called "nerve growth factor." "It would be inappropriate to suggest that this approach could be used to treat the course of normal aging in all organs, but it is not a far stretch to suggest that this may be useful in the treatment of Alzheimer's disease.

Other studies performed on various experimental model systems indicate that gene therapy can increase longevity and slow aging, even if in laboratories. Generally, such procedure required sophisticated technical processes. Overexpression of some genes, such as stress response and antioxidant genes, in some model systems also extends their longevity.

Overexpression of some genes increases life span in model systems. Human life span is associated with polymorphisms in genes.

Some such transgenic manipulations include the addition of gene(s) such as antioxidant genes superoxide dismutase (SOD) and catalase, NAD+-dependent histone deacetylases sirtuins, forkhead transcription factor FOXO, heat–shock proteins (HSP), heat–shock factor, protein repair methyltransferases and klotho, which is an inhibitor of insulin and IGF-1 signaling. Another system in which genetic interventions have been tested is the Hayflick system of limited proliferative life span of normal diploid differentiated cells in culture.

Leonard Hayflick, PhD, a professor of anatomy at the University of California, San Francisco is best known for his aging theory known as the Hayflick Limit, which places the maximum potential lifespan of humans at 120, the time at which too many cells can no longer split and divide to keep things going. This means that most of human cells stop dividing at age 120 years (repeated clause)

Hayflick said "Aging occurs because the complex biological molecules of which we are all composed become dysfunctional over time as the energy necessary to keep them structurally sound diminishes. Thus, our molecules must be repaired or replaced frequently by our own extensive repair systems." The question is that "can gene therapy be used to repair our aged molecules and cells"? Extensive studies showed that gene therapy can extend the life span of prokaryotic species.

Hayflick added in his July speech at the World Congress of Gerontology and Geriatric in Paris "These repair systems, which are also composed of complex molecules, eventually, suffer the same molecular dysfunction. The time when the balance shifts in favor of the accumulation of dysfunctional molecules is determined by natural selection — and leads to the manifestation of age changes that we recognize are characteristic of an old person or animal. It must occur after both reach reproductive maturity, otherwise the species would vanish." He continued to say "These fundamental molecular dysfunctional events lead to an increase in vulnerability to age-associated disease. Therefore, the study, and even the resolution of age-associated diseases, will tell us little about the fundamental processes of aging."

To conclude, Hayflick pointed out that Some types of cells, such as those that produce red and white blood corpuscles, can divide millions of times. Others, such as most nerve cells, do not reproduce at all. If a cell's Hayflick limit is 50, for example, it will divide 50 times and then become senescent. It withers and dies. When enough of our cells die, we die.

In 1961, Leonard Hayflick and Paul Moorhead discovered that human cells derived from embryonic tissues can only divide a finite number of times in culture. They divided the stages of cell culture in three phases: Phase I is the primary culture, when cells from the explants simply multiply to cover the surface of the culture flask. Phase II represents the period when cells divide in culture. Briefly, once cells cover a flask's surface, they stop multiplying. For cell growth to continue, the cells must be sub cultivated. To do so, one removes the culture's medium and adds a digestive enzyme called trypsin that dissolves the substances keeping cells together. If you add growth medium afterwards, you obtain the cells in suspension that can then be divided by two--or more--new flasks. Later, cells attach to the flask's floor and start dividing once again until new sub cultivation is required. Cells divide vigorously and can often be subcultivated in a matter of a few days. Eventually, however, cells start dividing slower, which marks the beginning of Phase III. Eventually they stop dividing at all and may or not die. Hayflick and Moorhead noticed that cultures stopped dividing after an average of fifty cumulative population doublings. This phenomenon is known as Hayflick's limit, Phase III phenomenon, or, as it will be called herein, replicative senescence as per Figure (85).

Figure (85): Phases of replication of cell culture

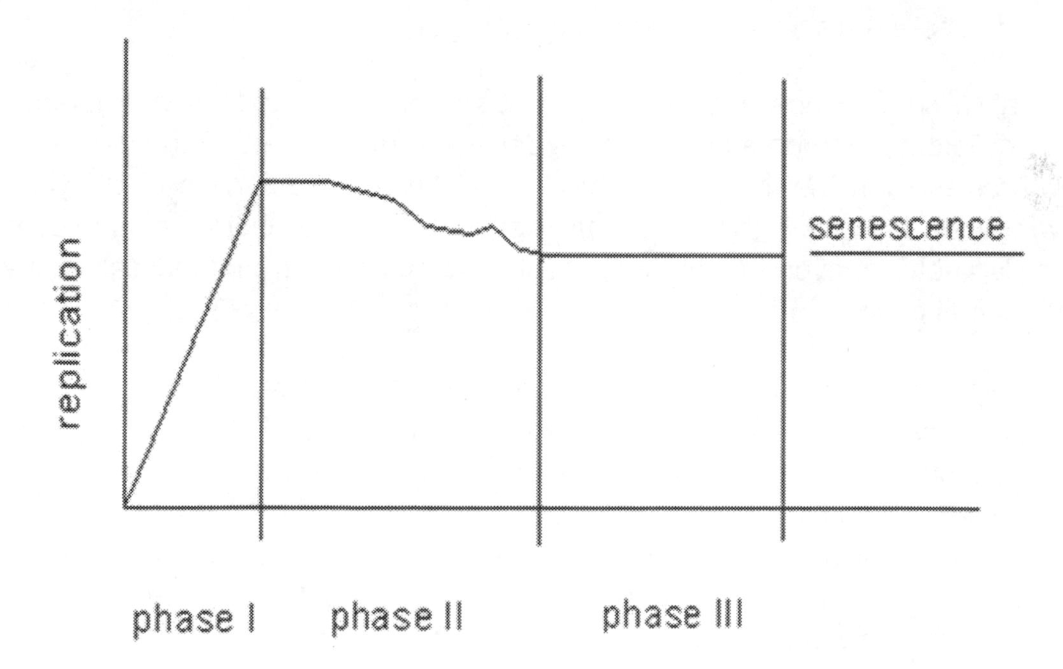


Hayflick and Moorhead worked with fibroblasts, a type of cell found in connective tissue; produces collagen, but RS (replicative senescence) has been found in other cell types: keratinocytes (epidermal cells), endothelial cells (endothelium cells are the thin layers of cells that line the interior surface of blood vessels), lymphocytes (a lymphocyte is a type of white blood cell in the immune system), adrenocortical cells, vascular smooth muscle cells, chondrocytes (are the only cells found in cartilage), etc. In addition, RS is observed in cells derived from embryonic tissues, in cells from adults of all ages, and in cells taken from many animals: mice, chickens, Galapagos tortoises, etc. Early results suggested a relation between the number of CPDs (cumulative population doublings) cells undergo in culture and the longevity of the species from which the cells were derived. For example, cells from the Galapagos tortoise, which can live over a century, divide about 110 times, while mouse cells divide roughly 15 times. In addition, cells taken from patients with progeroid syndromes such as Werner syndrome endure far less CPDs than normal cells. Exceptions exist and certain cell lines never reach RS. Some cells have no Hayflick limit. Barring trauma from outside, they are immortal. They can be killed, but they do not age. The "lowly" bacteria are immortal. They can be killed – by heat, starvation, radiation, lack of water, or being eaten by another organism. But they do not age. Bacteria keep on dividing forever, until some outside agency kills them. Cancer cells are similarly immortal. They keep on dividing and dividing, endlessly, unless they are killed or their host dies. "HeLa" cells, taken from the tumor of Henrietta Lacks in 1951 (Henrietta Lacks (August 18 (?), 1920 – October 4, 1951) was the involuntary donor of cells from her cancerous tumor, which were cultured by George Otto Gey to create an immortal cell line for medical research. This is now known as the Hela cell line), are still reproducing as vigorously as they did nearly 50 years ago. Human germline cells -- ova and sperm cells -- also show no Hayflick limit.

Replicative senescence (replicative CS) has been described for all metabolically active cells that undergo a spontaneous decline in growth rate. Senescent cells are growth arrested in the transition from phase G1 to phase S of the cell cycle. The growth arrest in RS is irreversible in the sense that growth factors cannot stimulate the cells to divide, even though senescent cells can remain metabolically active for long periods of time.

Recent studies showed that marrow cells (Marrow stromal cells, MSCs) may be used to repair senescence cells and gene therapy. Luk JM, Wang PP, Lee CK, et al. Hepatic potential of bone marrow stromal cells: development of in vitro co-culture and intra-portal transplantation models. J Immunol Method If marrow cells are to be used for cell and gene therapy, it will be important to define the conditions for isolation and expansion of the cells. As demonstrated by Friedenstein and colleagues, MSCs are relatively easy to isolate from marrow from most species by their adherence to tissue culture plates and flasks. However, the cells display several unusual features as they expand in culture. The difficulty of carrying out experiments in animal models with MSCs and other marrow cells has prompted scientists to develop a coculture system to study the repair of injured cells and tissues by MSCs. In initial experiments, MSCs were cocultured with heat-shocked human small airway epithelial cells. Figure (86) is a schematic showing MSCs with gene green fluorecent protein injected into senescence cells to produce repaired cells.

Figure (86): Using MSCs as gene therapy

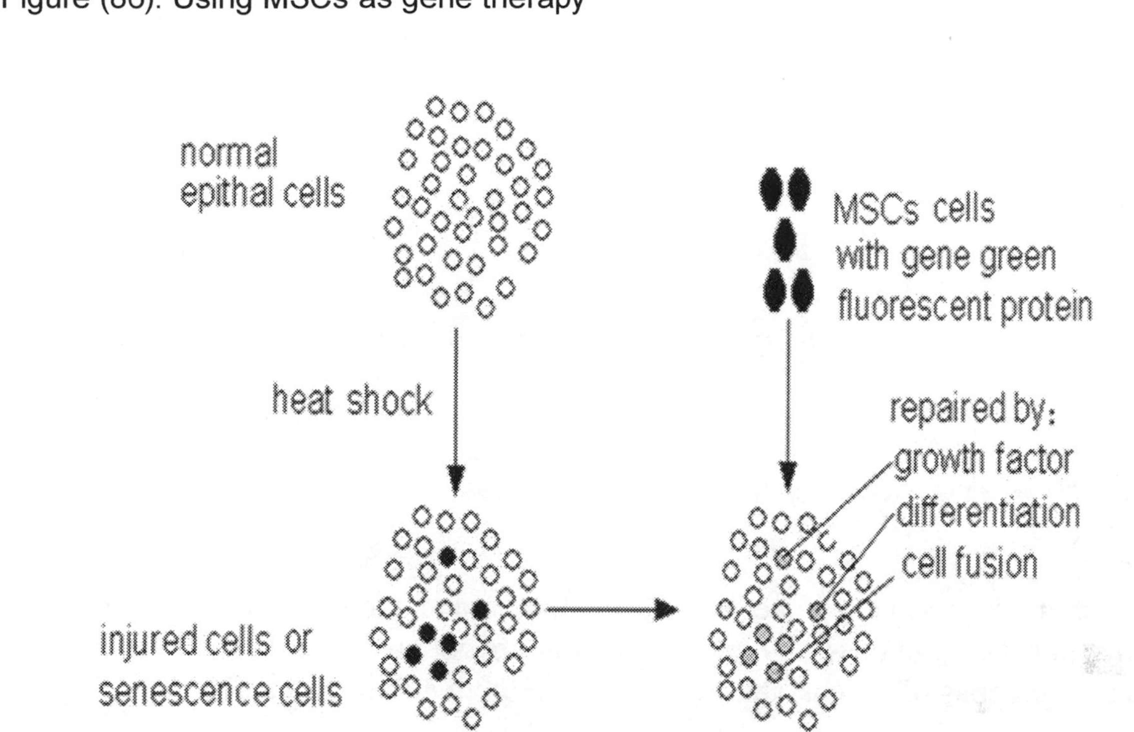

CHAPTER 7

# CELL BIOLOGY

Studying cells introduce you to the cell as the fundamental unit of life and the scientific method.

Mitosis is a form of eukaryotic cell division that produces two daughter cells with the same genetic component as the parent cell. Meiosis understands the events that occur in process of meiosis that takes place to produce our gametes.

Prokaryotes, Eukaryotes, and viruses learn about the cells that make up all living systems, their organelles, and the differences between living cells and viruses.

The cytoskeleton learns that the cytoskeleton acts both a muscle and a skeleton, and is responsible for cell movement, cytokinesis, and the organization of the organelles within the cell.

## 7.1 Cell Division

Living cells divide to form new cells in order to repair worn-out or damaged tissues throughout an organism, and (in the gametes only) to enable the exchange of genetic material at the initial stage of the process of sexual reproduction. In the human reproductive process, two kinds of sex cells, or gametes, are involved. The male gamete, or sperm, and the female gamete, the egg or ovum, meet in the female's reproductive system.

The two types of cell division are generally called mitosis and meiosis but, strictly, these terms refer to the stages of division of the cell nucleus for somatic (non-reproductive) and reproductive cells, respectively.

### 7.1.1 Mitosis

Eukaryotic cells reproduce in two different ways: mitosis and meiosis. In mitosis, the nucleus divides in two nuclei (referred to as karyokinesis), and cytoplasm divides in two portions (referred to as Cytokinesis), Figure (87).

Figure (87): Reproduction of two cells from one cell (mitosis)

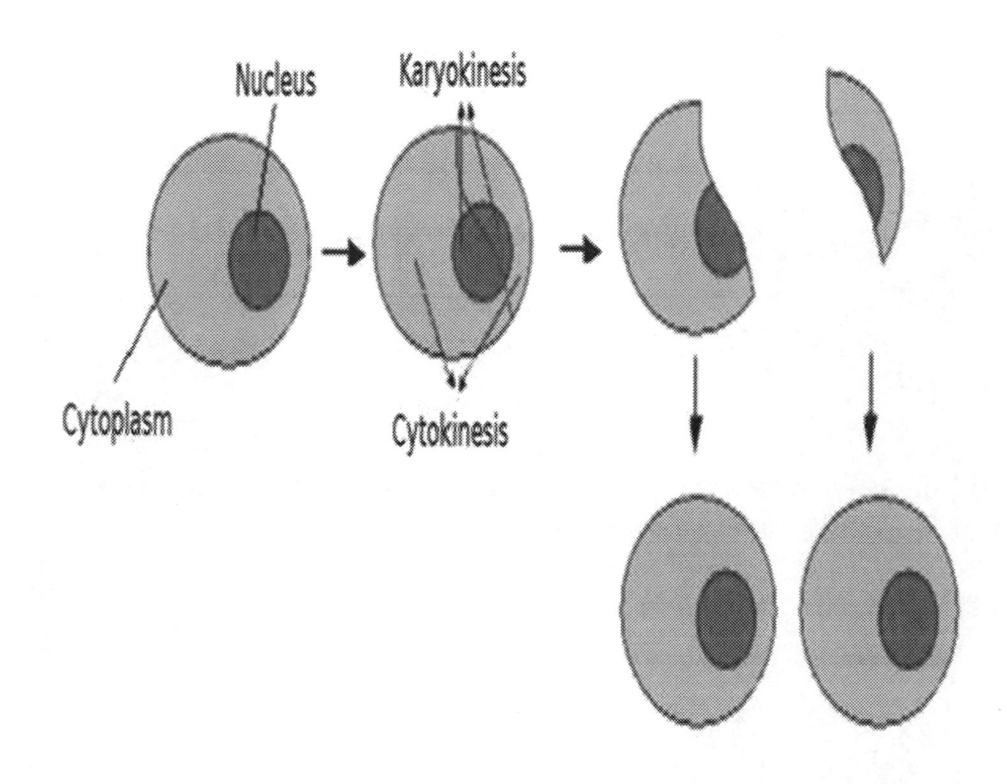


In the karyokinesis division, the chromosomes separate first and then are followed immediately by cytokinosis in which the cell divides into two identical daughters, Figure (88).

Figure (88): Division of nucleus

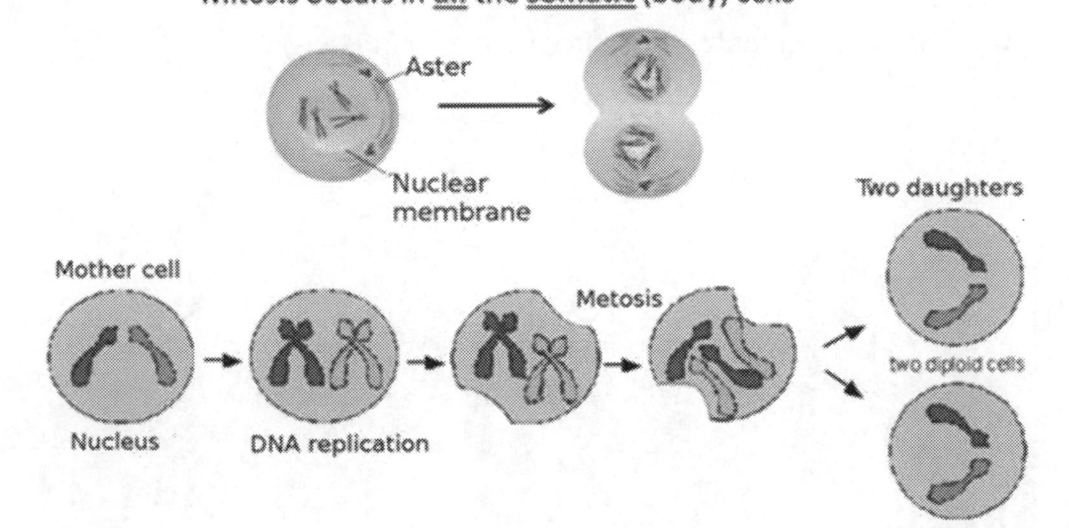


The period between successive divisions is known as interphase. The interphase period is divided into a number of stages, Figure (89).

Figure (89): Stages of cell division (mitosis)

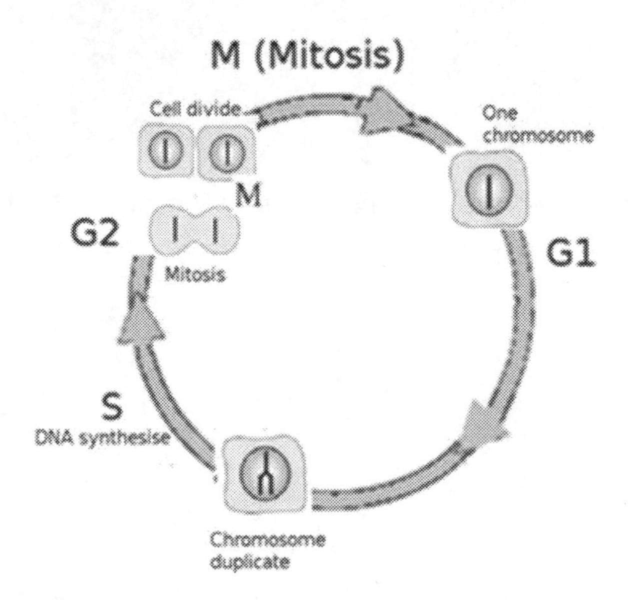


Some cells divide rapidly, for example, beans take about 19 hrs to complete the cycle division, whereas red blood cells divide at a rate of 2.5 million per second. Cancer cells divide rapidly such that the daughter cells divide before they have reached maturity.

Electrocharge, pH, temperature, and some drugs (enzymes) may affect the rate of division. When cells stop dividing, they stop at a point late in the stage G1. The stage S is the stage when the DNA is replicated for the next division, and the chromosomes become double stranded. The cell is then enters the G2 stage and proceeds into cell division. Cells will not divide again and stop in the G1. The process of division passes into the following steps:

- Prophase

Prophase is the first step of mitosis preparation where the cell is about to divide, and the chromosomes become visible and start to condense into double stranded chromosomes, Figure (90). Chromatin/DNA do not replicate in this step. Gradually, a spindle composed of protein fibers together with kinetochores form and extends nearly the length of the cell, expanded in its centre (or the equator of the cell) like a base ball. The chromosomes, each consisting of two chromotids attached to each other by a spindle fiber at the centromere.

Figure (90): Double stranded chromosome divides into two identical sisters

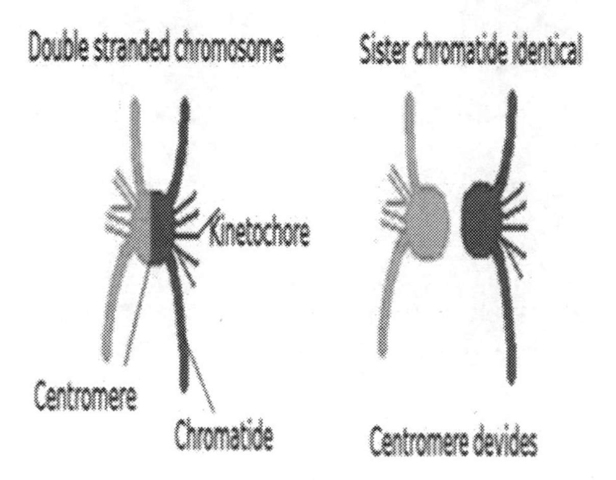


- Metaphase

When the centomere and kinetochore arrive at the centre of the cell (equator), the metaphase begins. The mitosis process ends when the centromere divides so that each of the chromatids becomes a single stranded chromosome, Figure (63).

- Anaphase

During anaphase, each sister of a single-stranded chromosome moves towards one pole of the cell (one sister to one pole, and the other to the opposite pole). Thus, anaphase is the stage when sisters of chromosome migrate to the two poles of the cell, Figure (91).

Figure (91): Movement of daughter chromosomes from equator to poles during anaphase

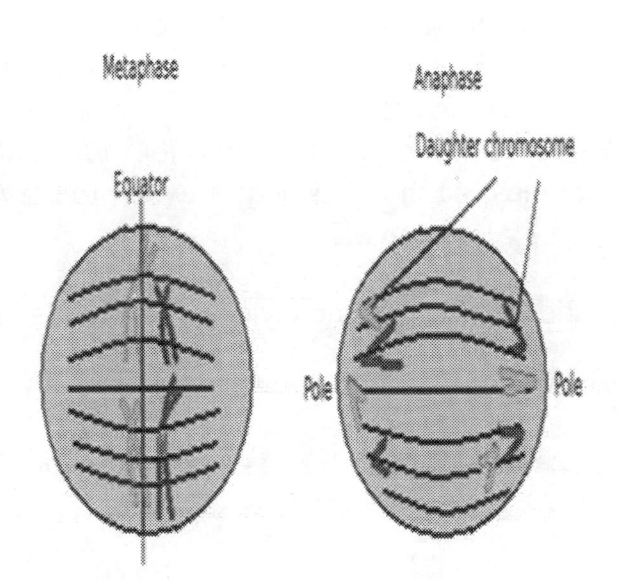


- Telophase

In telophase, the nuclear envelope (which had disappeared during prophase) reforms, and chromosomes uncoil into chromatin form again. There are now two cells instead of one cell, but of smaller sizes of same genetic heredity. The small cells will develop into mature ones.

- Cytokinesis

Cytokinesis is the last stage of cell division, where the daughter cells split apart. The mitosis is the division of the nucleus, and the cytokinesis is the division of the cytoplasm.

### 7.1.2 Meiosis

During meiosis chromosomes are duplicated once in S stage. Meiosis has two stages; the meiosis 1 is just like mitosis where the cell divides once. In meiosis 2 the cell again divides and the final product is four daughter cells. The process in meiosis is as follows:

- The first step in meiosis is the diploids of each of the 46 chromosomes duplicated, Figure (65). This is called DNA replication.
- The homologues of chromosomes (23 homologues in 46 chromosomes; i.e. one chromosome has two homologues) are segregated in two cells.
- Homologues pair up alongside lengthwise, no.4 of Figure (92).
- The pairs of homologues are shuffling and cross over in metaphase 1, figure (48).
- At the end of meiosis, new chromosomes will be created to become part of eggs and sperms that will be unique to their parent.
- During meiosis 2, the sister chromatids of each of the 23 chromosomes are pulled apart resulting in 4 cells that each of the 23 chromosomes are now haploid.
- Meiosis process is the way to create the diversity of all sexually reproducing organisms.
- The invention of electronic microscopes allowed biologists to discover the basic facts of cell division and sexual reproduction. The center of genetics and heredity research then shifted to understanding what really happens in the transmission of hereditary traits from parents to children. A number of hypotheses were suggested by researchers but Gregor Mendel, a monk from Austria was the father of heredity.

Figure (92): Stages of meiosis

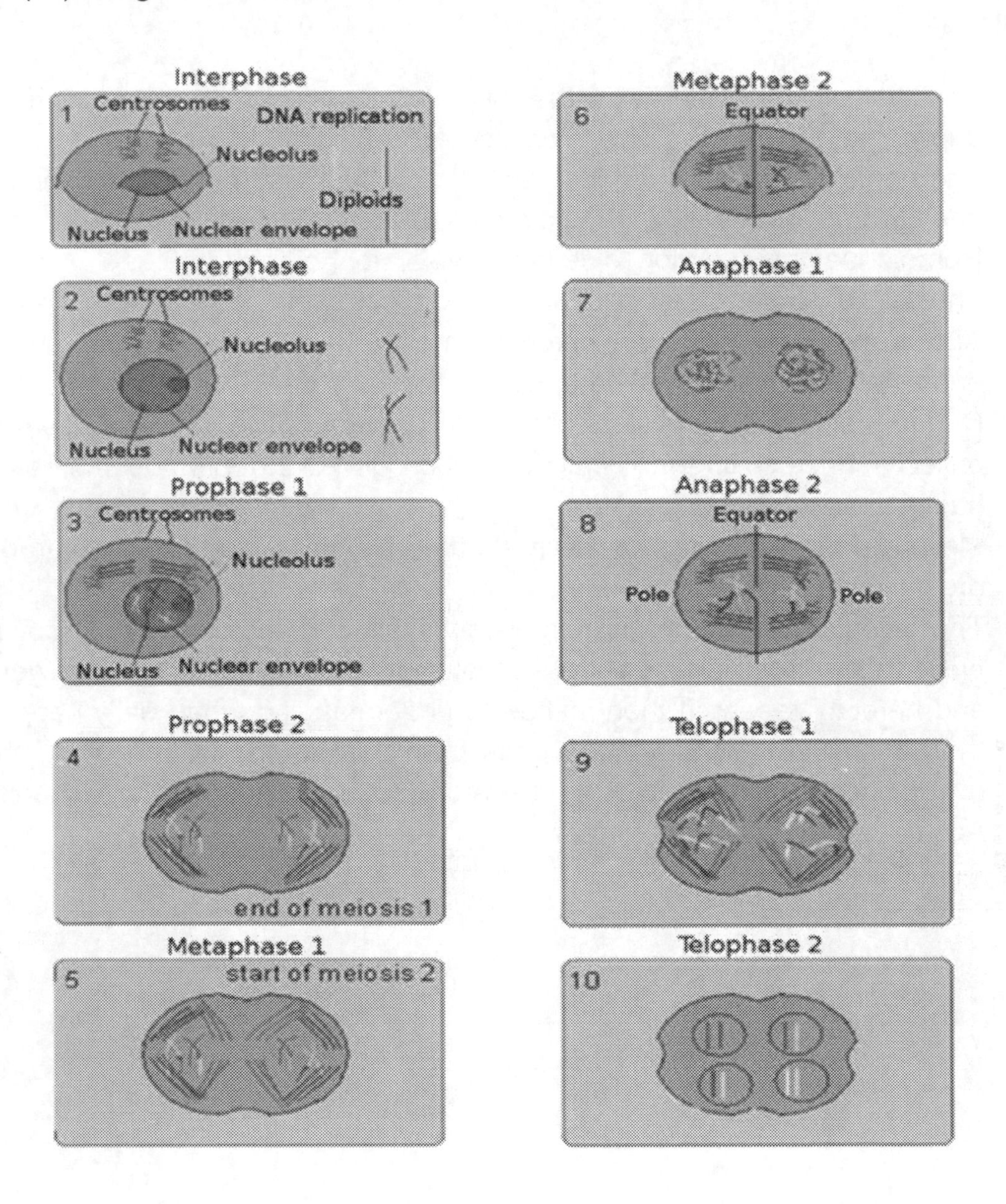


### 7.1.3 Why Meiosis?

Researchers at the University of California, Davis have discovered a key tool that helps sperm and eggs develop exactly 23 chromosomes each. The work, which could lead to insights into fertility, spontaneous miscarriages, cancer and developmental disorders, is published April 13 in the journal *Cell*.

Healthy humans have 46 chromosomes, 23 from the sperm and 23 from the egg. An embryo with the wrong number of chromosomes is usually miscarried, or developed in disorders such as Down's syndrome, which is caused by an additional copy of chromosome 21.

During meiosis, the cell division process that creates sperm and eggs, matching chromosomes pair up and become connected by “crossing over” with each other, said Neil Hunter, a professor of microbiology at UC Davis and senior author of the new study.

These connections are essential for precise chromosome sorting and the formation of sperm and eggs with exactly the right numbers of chromosomes. Crossovers also play a fundamental role in evolution by allowing the chromosomes to swap chunks of DNA, introducing some variety into the next generation. Crossing over was described, in theory, by Thomas Hunt Morgan. He relied on the discovery of the Belgian Professor Frans Alfons Janssens of the University of Leuven who described the phenomenon in 1909 and had called it “chiasmatypie”.

Each pair of chromosomes must contain at least one crossover. But there shouldn’t be more than about two crossovers per pair, or the genome could be destabilized.

In their paper, Hunter and his colleagues describe a “missing tool” that explains how crossovers are regulated.

“There must be enzymes that ensure at least one crossover, but not too many,” said Hunter, who is also a member of the UC Davis Comprehensive Cancer Center research program.

Hunter, graduate students Kseniya Zakharyevich and Shangming Tang and research associate Yunmei Ma, looked for enzymes that could cut DNA to form crossovers in yeast, which form sexual gametes, or spores, in much the same way that humans and other mammals form sperm and eggs.

“There were several good candidates, but none turned out to play a major role,” Hunter said.

Then they discovered the missing tool for crossing-over: three yeast enzymes, Mlh1, Mlh3 and Sgs1, which work together to cut DNA and make crossovers.

It turns out that the human equivalents of these enzymes are well known for their role in suppressing tumors. Human MLH1 and MLH3 are mutated in an inherited form of colon cancer. BLM, the human equivalent of Sgs1, is mutated in a cancer-prone disease called Bloom’s Syndrome.

“Sgs1 was the biggest surprise,” Hunter said. “We previously knew it as an enzyme that unwinds DNA to prevent crossovers. Its role in making crossovers had been hidden by other enzymes that can step in when it is absent.”

"While other enzymes cut DNA randomly, Mlh1-Mlh3-Sgs1 only makes crossovers. This unique activity is essential for meiosis and its discovery is a huge step forward," he said.

7.1.4 Alternation of Generations

Life cycles of plants and algae with alternating haploid and diploid multicellular stages are referred to as diplohaplontic (the equivalent terms haplodipontic, diplobiontic or dibiontic are also in use). Life cycles, such as those of animals, in which there is only a diploid multicellular stage are referred to as diplontic. (Life cycles in which there is only a haploid multicellular stage are referred to as haplontic.). It turns out plants reproduce both sexually (sperm and eggs) and asexually (spores), and both types of reproduction are necessary to complete the cycle. That is, a sexually reproducing plant will make a plant that reproduces asexually, and then the asexually reproducing plant will make a sexually reproducing plant, and the circle goes round and round. Each reproductive event produces a new generation, and they alternate the types of reproduction, thus the name alternation of generations. What's a generation? You and your siblings are one generation, and your parents are a different (and older) generation, and your grandparents are another, even older, generation. There are special names for each of the plants we've been talking about. A sporophyte is a plant that produces spores (asexual reproduction) and a gametophyte is a plant that produces gametes (sexual reproduction).

Asexual reproduction is the primary form of reproduction for single-celled organism as the archaebacteria, eubacteria, and protists. Many plants and fungi reproduce asexually as well. In animals, sexual reproduction involves the two alternating processes of meiosis and fertilization.

- In meiosis, the chromosome number is reduced from the diploid to the haploid number.
- In fertilization, the nuclei of two gametes fuse, raising the chromosome number from haploid to diploid. Whatever variation in details there may be from one organism to another, these two activities must occur alternately if sexual reproduction is to continue, Figure (93).

In animals, meiosis generates the haploid gametes — sperm and eggs — directly. These single cells fuse to form the zygote which will develop into another diploid animal.

In plant life cycles, there are two alternating phases, a diploid (2N) phase and a haploid (N) phase, known as alternation of generations. During the two phases of the life cycle, shown in the figure below, mitosis and meiosis alternate to produce the two types of reproductive cells—gametes and spores. The diploid (2N) phase is known as the sporophyte, or spore-producing plant. The haploid (N) phase is known as the gametophyte, or gamete-producing plant. Plant spores are haploid (N) reproductive cells formed in the sporophyte plant by meiosis that can grow into new individuals. The new individual is the gametophyte. A gamete is a reproductive cell that is produced by mitosis and fuses during fertilization with another gamete to produce a new individual, the diploid sporophyte.

All plants have a life cycle with alternation of generations, in which the haploid gametophyte phase alternates with the diploid sporophyte phase. In plants, meiosis results in the production of spores, not gametes.

Figure (93): Animal life cycle

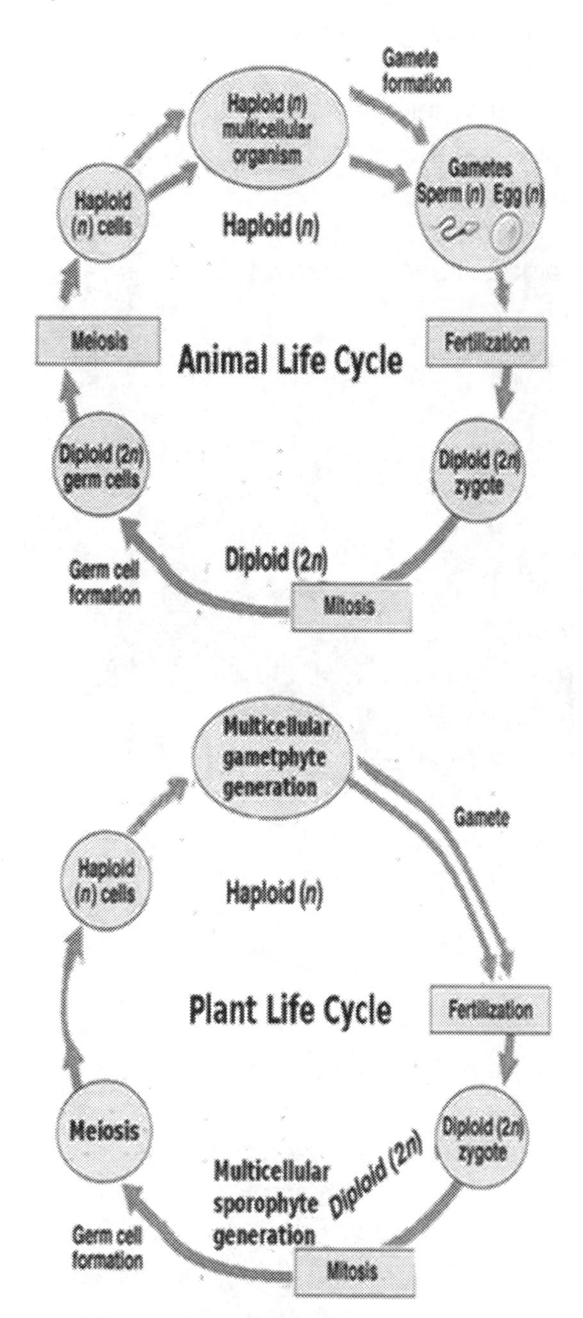

### 7.1.5 Biological life cycle

A life cycle is a period involving one generation of an organism through means of reproduction, whether through asexual reproduction or sexual reproduction. In regard to its ploidy, there are three types of cycles; haplontic life cycle, diplontic life cycle, diplobiontic life cycle. These three types of cycles feature alternating haploid and diploid phases (n and 2n). The haploid organism becomes diploid through fertilization, which joins of gametes. This results in a zygote which then germinates. To return to a haploid stage, meiosis must occur. The cycles differ in the product of meiosis, and whether mitosis (growth) occurs. Zygotic and gametic meioses have one mitotic stage and form: during the n phase in zygotic meiosis and during the 2n phase in gametic meiosis.

Therefore, zygotic and gametic meiosis are collectively term haplobiontic (single mitosis per phase). Sporic meiosis, on the other hand, has two mitosis events (diplobiontic): one in each phase.

The life cycle of green algae is shown below in Figure (94).

Figure (94): Algae life cycle

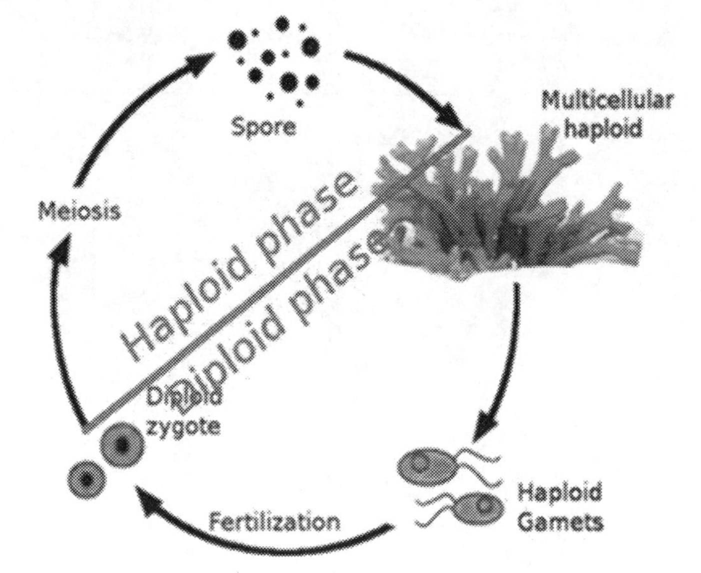


Haploid spores give rise to a multicellular haploid leaf-like structure called a thallus. The thallus produces gametes. Green algae are isogamus, meaning they have only one type of gamete, rather than having separate male and female gametes. When two gametes meet, fertilization takes place and a diploid zygote is formed. The zygote then germinates, undergoes meiosis and forms haploid spores. The diploid phase of the life cycle is brief and unicellular. There are a few exceptions this general life cycle, such as theUlva (sea lettuce), which has a multicellular diploid phase similar to that found in brown algae.

Life cycle can be changed, depends on the environment and the type of the host. Different hosts are shown in Figure (95).

Figure (95): Different hosts for ticks

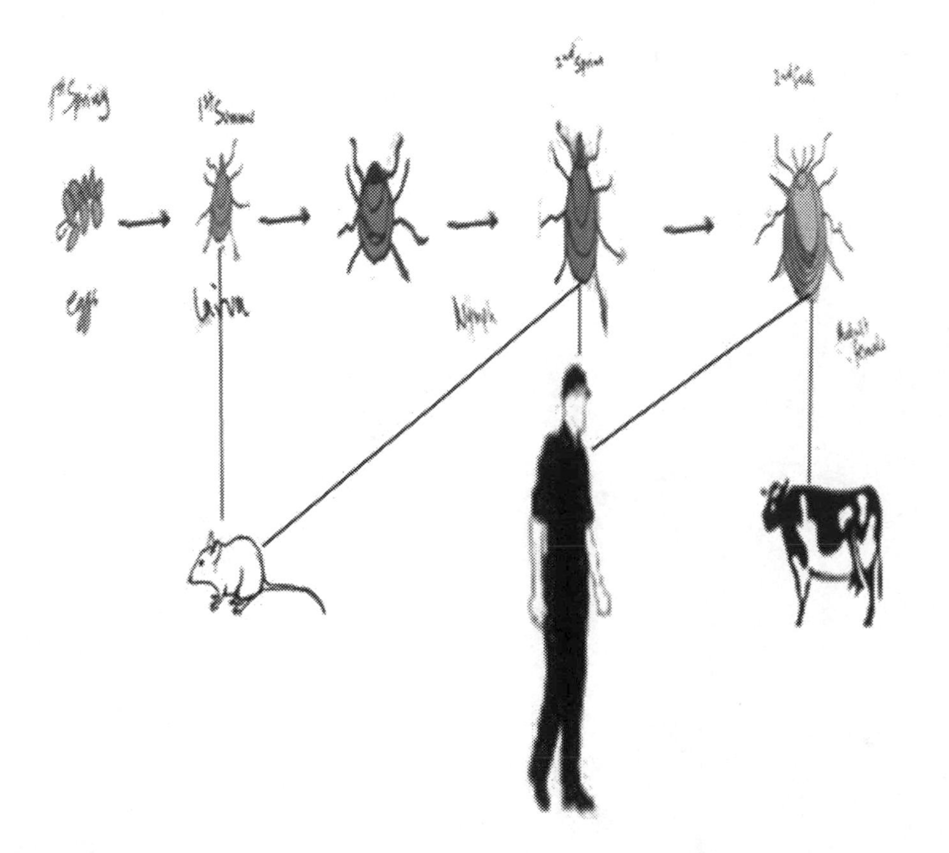

### 7.1.5.1 Parthenogenesis

Several methods have been used for activation of oocytes after ICSI such as electrical activation, mechanical activation, and chemical activation such as ethanol, ionophores, strontium and 6-dimethylaminopurine Using chemical or electrical stimuli, it is also possible to stimulate human eggs to undergo several rounds of cell division, as if they had been fertilized. In this case, the egg retains all forty-six egg cell chromosomes and egg cell mitochondria. In amphibians, this asexual reproduction process, known as parthenogenesis, has produced live offspring that contain the same nuclear DNA as the egg. These offspring are all necessarily female. Parthenogenesis in mammals has not led reproducibly to the production of live offspring. Rougier, N., and Z. Werb. “Minireview: Parthenogenesis in mammals” *Mol Reprod Devel* 59: 468-474, 2001.

Cibelli et al. (Cibelli, J.B., et al. “Somatic cell nuclear transfer in humans: Pronuclear and early embryonic development” ebiomed: *The Journal of Regenerative Medicine*,

2: 25-31, 2001) activated human eggs (obtained from informed and consenting donors) by parthenogenesis, and obtained multiple cell divisions up to the early embryo stage in six out of twenty-two attempts. Although there was no report that stem cells were isolated in these experiments, it is possible that parthenogenesis of human eggs could induce them to develop to a stage where parthenogenetic stem cells could be isolated. For example, Cibelli et al. derived a monkey parthenogenetic stem cell preparation from Macaca fasicularis eggs activated by parthenogenesis. Whether cloned stem cells resulting from parthenogenesis have been completely and correctly epigenetically reprogrammed remains to be determined, Figure (96).

Figure (96): Normal fertilization, cloning and parthenogenesis

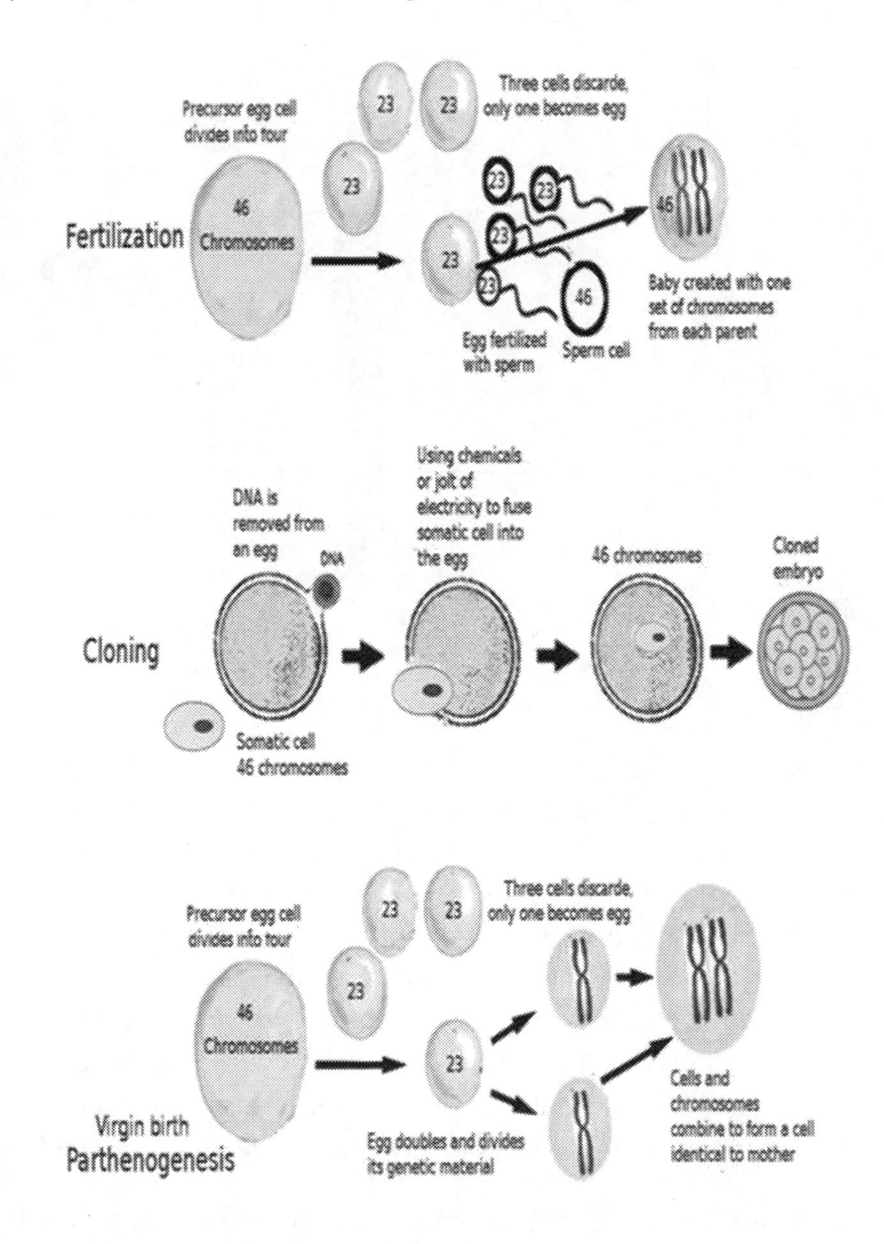

Parthenogenesis, in which an unfertilized egg develops to maturity, has been found in 70 species of vertebrates, including captive snakes and a monitor lizard species.

## 7.2 Bacterial Conjugation

Bacterial conjugation is often regarded as the bacterial equivalent of sexual reproduction or mating since it involves the exchange of genetic material. During conjugation the donor cell provides a conjugative or mobilizable genetic element that is most often a plasmid or transposon. Most conjugative plasmids have systems ensuring that the recipient cell does not already contain a similar element. Conjugation is the direct transfer of DNA from one bacterial cell to another bacterial cell. The transferred DNA is a plasmid, a circle of DNA that is distinct from the main bacterial chromosome. The F plasmid (F stands for fertility) is similar to a virus or a transposon in its ability to move independently of the main chromosome. One strand of the plasmid is transferred and the other remains in the original cell. Both strands have the complementary stranded added so that each cell ends up with a complete plasmid.

Bacterial conjugation is often regarded as the bacterial equivalent of sexual reproduction or mating since it involves the exchange of genetic material. During conjugation the *donor* cell provides a conjugative or mobilizable genetic element that is most often a plasmid or transposon (A transposable element is a DNA sequence that can change its position within a genome, sometimes creating or reversing mutations and altering the cell's genome size). Most conjugative plasmids have systems ensuring that the recipient cell does not already contain a similar element.

The genetic information transferred is often beneficial to the recipient. Benefits may include antibiotic resistance, xenobiotic tolerance or the ability to use new metabolities. Such beneficial plasmids may be considered bacterial endosymbionts. Other elements, however, may be viewed as bacterial paraites and conjugation as a mechanism evolved by them to allow for their spread.

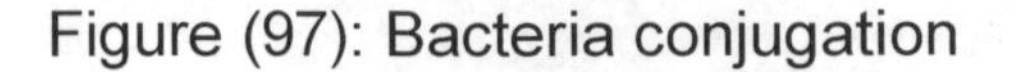

Figure (97): Bacteria conjugation

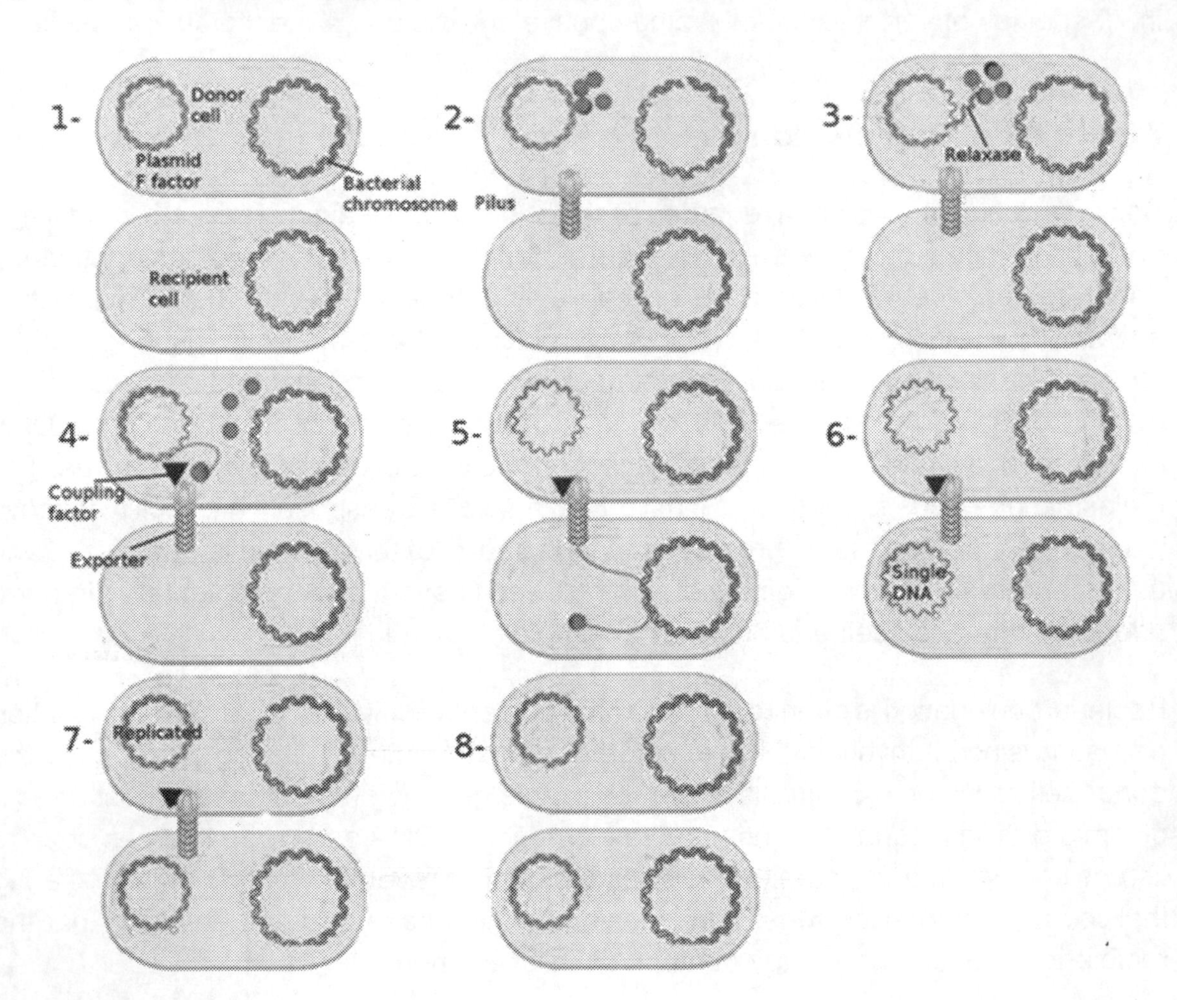


## 7.3 DNA Fingerprinting

DNA profiling (also called DNA fingerprinting, DNA testing, or DNA typing) is a forensic technique used to identify individuals by characteristics of their DNA. A DNA profile is a small set of DNA variations that is very likely to be different in all unrelated individuals, thereby being as unique to individuals as are fingerprints (hence the alternate name for the technique). DNA profiling should not be confused with full genome sequencing. First developed and used in 1985. DNA fingerprinting**,** also called DNA typing, DNA profiling, genetic fingerprinting, genotyping, or identity testing, in genetics, method of isolating and identifying variable elements within the base-pair sequence of DNA (deoxyribonucleic acid). The technique was developed in 1984 by British geneticist Alec Jeffreys, after he noticed that certain sequences of highly variable DNA (known as minisatellites), which do not contribute to the functions of genes, are repeated within genes. Jeffreys recognized that each individual has a unique pattern of minisatellites (the only exceptions being multiple individuals from a single zygote, such as identical twins).

DNA (deoxyribonucleic acid) represents the blueprint of the human genetic makeup. It exists in virtually every cell of the human body and differs in its sequence of nucleotides (molecules that make up DNA, also abbreviated by letters, A, T, G, C; or, adenine, thymine, guanine, and cytosine, respectively). The human genome is made up of 3 billion nucleotides, which are 99.9% identical from one person to the next. The 0.1% variation, therefore, can be used to distinguish one individual from another. It is this difference that can be used by forensic scientists to match specimens of blood, tissue, or hair follicles to an individual with a high level of certainty. The procedure for creating a DNA fingerprint consists of first obtaining a sample of cells, such as skin, hair, or blood cells, which contain DNA. The DNA is extracted from the cells and purified. In Jeffreys's original approach, which was based on restriction fragment length polymorphism (RFLP) technology, the DNA was then cut at specific points along the strand with proteins known as restriction. The enzymes produced fragments of varying lengths that were sorted by placing them on a gel and then subjecting the gel to an electric current (electrophoresis): the shorter the fragment, the more quickly it moved toward the positive pole (anode). The sorted double-stranded DNA fragments were then subjected to a blotting technique in which they were split into single strands and transferred to a nylon sheet. The fragments underwent autoradiography in which they were exposed to DNA probes—pieces of synthetic DNA that were made radioactive and that bound to the minisatellites. A piece of X-ray film was then exposed to the fragments, and a dark mark was produced at any point where a radioactive probe had become attached. The resultant pattern of marks could then be analyzed.

## 7.4 Genetic Relationship between Individuals

Genetic genealogy is the use of DNA testing in combination with traditional genealogy and traditional genealogical and historical records to infer relationships between individuals. Genetic genealogy involves the use of genealogical DNA testing to determine the level and type of the genetic relationship between individuals. This application of genetics became popular with family historians in the first decade of the 21st century, as tests became affordable. The tests have been promoted by amateur groups, such as surname study groups, or regional genealogical groups, as well as research projects such as the genographic project.. Genetic genealogy is the use of DNA testing in combination with traditional genealogy and... As of 2013 hundreds of thousands of people had been tested. As this field has developed, the aims of practitioners broadened, with many seeking knowledge of their ancestry beyond the recent centuries for which traditional pedigrees can be constructed.

The investigation of surnames in genetics can be said to go back to George Darwin, a son of Charles Darwin. In 1875, George Darwin used surnames to estimate the frequency of first-cousin marriages and calculated the expected incidence of marriage between people of the same surname (isonomy). He arrived at a figure between 2.25% and 4.5% for cousin-marriage in the population of Great Britain, higher among the upper classes and lower among the general rural population.

### 7.4.1 Surname studies

One famous study examined the lineage of descendants of Thomas Jefferson's paternal line and male lineage descendants of the freed slave, Sally Hammings.

Bryan Sykes, a molecular biologist at Oxford University tested the new methodology in general surname research. His study of the Sykes surname obtained results by looking at four STR markers on the male chromosome. It pointed the way to genetics becoming a valuable assistant in the service of genealogy and history.

### 7.4.2 Direct to consumer paternity testing

The first company to provide direct-to-consumer genetic DNA testing was the now defunct Gene Tree. However, it did not offer multi-generational genealogy tests. In fall 2001, GeneTree sold its assets to Salt Lake City-based Sorenson Molecular Genealogy Foundation (SMGF) which originated in 1999.[4] While in operation, SMGF provided free Y-Chromosome and mitochondrial DNA tests to thousands.[5] Later, GeneTree returned to genetic testing for genealogy in conjunction with the Sorenson parent company and eventually was part of the assets acquired in the Ancestry.com buyout of SMGF.

### 7.4.3 The genetic genealogy revolution

In 2000, Family Tree DNA, founded by Bernett Greenspan and Max Blankfeld, was the first company dedicated to direct-to-consumer testing for genealogy research. They initially offered eleven marker Y-Chromosome STR tests and HVR1 mitochondrial DNA tests. They originally tested in partnership with the University of Arizona.

The publication of The Seven Daughters of Eve by Sykes in 2001, which described the seven major haplogroups of European ancestors, helped push personal ancestry testing through DNA tests into wide public notice. With the growing availability and affordability of genealogical DNA testing, genetic genealogy as a field grew rapidly. By 2003, the field of DNA testing of surnames was declared officially to have "arrived" in an article by Jobling and Tyler-Smith in *Nature Reviews Genetics*.[14] The number of firms offering tests, and the number of consumers ordering them, rose dramatically.

## 7.5 The Genographic Project

The original Genographic Project was a five-year research study launched in 2005 by the National Geographic Society and IBM, in partnership with the University of Arizona and Family Tree DNA. Its goals were primarily anthropological. The project announced that by April 2010 it had sold more than 350,000 of its public participation testing kits, which test the general public for either twelve STR markers on the Y-chromosome or mutations on the HVR1 region of the mtDNA.

In 2007, annual sales of genetic genealogical tests for all companies, including the laboratories that support them, were estimated to be in the area of $60 million (2006).

## 7.6 Human Genome Project

The Human Genome Project (HGP) was one of the great feats of exploration in history - an inward voyage of discovery rather than an outward exploration of the planet or the cosmos; an international research effort to sequence and map all of the genes - together known as the genome - of members of our species, *Homo sapiens.* Completed in April 2003, the HGP gave us the ability, for the first time, to read nature's complete genetic blueprint for building a human being.

The Human Genome Project (HGP) is an international scientific research project with the goal of determining the sequence of chemical base pairs which make up human DNA, and of identifying and mapping all of the genes of the human genome from both a physical and functional standpoint. It remains the world's largest collaborative biological project. After the idea was picked up in 1984 by the US government the planning started, with the project formally launched in 1990, and finally declared complete in 2003. Funding came from the US government through the National Institutes of Health as well as numerous other groups from around the world. A parallel project was conducted outside of government by the Celera Corporation, or Celera Genomics, which was formally launched in 1998. Most of the government-sponsored sequencing was performed in twenty universities and research centers in the United States, the United Kingdom, Japan, France, Germany, and China.

The Human Genome Project originally aimed to map the nucleotides contained in a human haploid reference genome (more than three billion). The "genome" of any given individual is unique; mapping "the human genome" involves sequencing multiple variations of each gene.

Key findings of the draft (2001) and complete (2004) genome sequences include:

1. There are approximately 20,500 genes in human beings, the same range as in mice.
2. The human genome has significantly more segmental duplications (nearly identical, repeated sections of DNA) than had been previously suspected.
3. At the time when the draft sequence was published fewer than 7% of protein families appeared to be vertebrate specific.

In this section, you will find access to a wealth of information on the history of the HGP, its progress, cast of characters and future, Educational Resources

- General Information
- Research
- Model Organism

A Quarter Century after the Human Genome Project's Launch: Lessons Beyond the Base Pairs

The National Human Genome Research Institute began as the National Center for Human Genome Research (NCHGR), which was established in 1989 to carry out the role of the National Institutes of Health (NIH) in the International Human Genome Project (HGP). October 1, 2015 marked the 25$^{th}$ anniversary of the launch of the Human Genome Project. To commemorate this anniversary the NHGRI History of Genomics Program is hosting a seminar series entitled "A Quarter Century after the Human Genome Project's Launch: Lessons beyond the Base Pairs."

## 7.11 Frozen Ark

The Frozen Ark is a charitable frozen zoo project created jointly by the Zoological Society of London, the Natural History Museum and University of Nottingham. The project aims to preserve the DNA and living cells of endangered to retain the genetic knowledge for the future. The Frozen Ark collects and stores samples taken from animals in zoos and those threatened with extinction in the wild, with the expectation that, some day, cloning technologies will have matured sufficiently to resurrect extinct species.[3] The Frozen Ark was a finalist for the Saatchi and Saatchi Award for World Changing Ideas in 2006.

# Glossary

Allele: One of two or more alternative forms of a particular gene that arise by mutation at a specific position on a chromosome.

Allelic heterogeneity: The phenomenon in which Different mutations in the same gene can cause a similar phenotype. Allelic heterogeneity can arise as a result of natural selection processes, as a result of exogenous mutagens, genetic drift, or genetic migration.

Alternative splicing: It is a regulated process during gene expression that results in a single gene coding for multiple proteins. The process where the initial strand of RNA copied directly from DNA is cut up (spliced) and pieced together into different messages (mRNAs) and can lead to different Proteins

Amino acid: It is a simple organic compound containing both a carboxyl (—COOH) and an amino (—$NH_2$) group. Amino acids play central roles both as building blocks of proteins and as intermediates in metabolism. The 20 amino acids that are found within proteins convey a vast array of chemical versatility. Our bodies make 12 and we get the other 8 (the 8 essential amino acids) from our food.

Amniotic fluid: It is, commonly called a pregnant woman's water, is the protective liquid contained by the amniotic sac of a pregnant female. It is mostly water but also has cells sloughed off from the fetus, secretions from the placenta and fetal urine.

Apoptosis: If cells are no longer needed, they commit suicide by activating an intracellular death program. This process is therefore called programmed cell death programmed cell death,. Similarly, while cellular senescence is shown to be an alternative to apoptosis, blocking damaged cells from proliferating, it is also capable of promoting tumorigenesis and aging.

Autosome: An autosome is any chromosome that is not a sex-determining chromosome, so most chromosomes are autosomes. Human beings have 22 pairs of autosomes and two sex chromosomes, XY for male and two Xs for female.

Bacterium: Bacteria are widely distributed in soil, water, and air, and on or in the tissues of plants and animals. Bacterium is a member of a large group of unicellular microorganisms that have cell walls but lack organelles and an organized nucleus.

Base pair: A base is the variable component of a nucleic acid. The rules of base pairing (or nucleotide pairing) are: A with T: the purine adenine (A) always pairs with the pyrimidine thymine (T) C with G: the pyrimidine cytosine (C) always pairs with the purine guanine (G)

Cancer: Cancer is a class of diseases characterized by out-of-control cell growth. There are over 100 different types of cancer, and each is classified by the type of cell that is initially affected.

Candidate gene: The gene may be a candidate because it is located in a particular chromosome region suspected of being involved in the disease or its protein product may suggest that it could be the disease gene in question. For example, the gene coding for a protein that works in the brain and regulates neuron transmitters may be a good candidate for causing depression.

Carrier (of a genetic disease): A carrier is a person who has a change in one copy of a gene. The carrier does not have the genetic disease related to the abnormal gene. Someone who has a disease-causing mutation in their DNA but does not show any symptoms.

Cas: It is for gene editing - CRISPR-associated genes

Cas9, Csn1: A CRISPR-associated protein containing two nuclease domains; Cas9 and Cas1

crDNA: RNA for CRISPR

Cell: The cell is the basic structural, functional, and biological unit of all known living organisms. It is the smallest unit of life that can exist independently. All organisms are made up of one or more cells

Centromere: The point on a chromosome by which it is attached to a spindle fiber during cell division (mitosis or meiosis).

Chromatin: It is a complex of DNA and proteins that forms chromosomes within the nucleus of eukaryotic cells. It is highly condensed and wrapped around nuclear proteins in order to fit inside the nucleus.

Chromosome: It is a threadlike structure of nucleic acids and protein found in the nucleus of most living cells, carrying genetic information in the form of genes. A long, twisted and folded-up piece of DNA. Humans have 46, 23 contributed by each parent.

Clone: It is the process of producing similar populations of genetically identical individuals that occurs in nature when organisms such as bacteria, insects or plants reproduce asexually. An organism, either single celled or multicellular, that has the exact same DNA as another organism.

Codominant: Two different versions of a gene contribute to the final trait so that neither version is masked by the other. When alleles for both white and red are present in a carnation, for example, the result is a pink carnation since both alleles are codominant.

Codon: It is a sequence of three DNA or RNA nucleotides that corresponds with a specific amino acid or stop signal during protein synthesis. Three adjacent bases (letters) that "code" for a particular amino acid in protein translation. There are 64 possible 3-letter combinations of the bases but only 20 amino acids, so several of the codons code for the same amino acid.

Complementary: A property of nucleic acids, whereby adenine (A) always pairs with thymine (T) while cytosine (C) always pairs with guanine (G). It is also double-stranded DNA synthesized from a single stranded RNA (e.g., messenger RNA (mRNA) or microRNA (microRNA)) template in a reaction catalysed by the enzyme reverse transcriptase

Congenital: (especially of a disease or physical abnormality) present from birth. The congenital trait can be due to genetic factors, environmental ones or a mixture of both.

CRISPR-Cas9: CRISPR stands for clustered regularly interspaced short palindromic repeat and is a customizable tool that lets scientists cut and insert These repeats were initially discovered in the 1980s in *E. coli* (9), but their function wasn't confirmed until 2007

Crossing over: Crossing over occurs in the first division of meiosis. Two members of a chromosome pair twist around one another and exchange genetic information. It is a process in genetics by which the two chromosomes of a homologous pair exchange equal segments with each other.

Diploid: The characteristic of an organism or cell having two complete sets of chromosomes in each cell. An example of a cell in a diploid state is a somatic cell. In humans, the somatic cells typically contain 46 chromosomes in contrast to human haploid gamets (egg and sperm cells) that have only 23 chromosomes.

DNA: DNA stands for deoxyribonucleic acid. It is the molecule that contains the genetic instructions to construct and maintain a living organism. DNA is the hereditary material in humans and almost all other organisms. Nearly every cell in a person's body has the same DNA.

DNA fingerprint: A technique used for identification, for example DNA fingerprinting is a laboratory technique used to establish a link between biological evidence and a suspect in a criminal investigation. Relatively large amounts of DNA are isolated and cut by restriction enzymes. Note: this technique has been essentially replaced by other types of DNA profiling.

DNA methylation: A biochemical process involving the addition of a methyl group to the cytosine of DNA nucleotides. When located in a gene promoter, DNA methylation typically acts to repress gene transcription.

DNA profiling: It is the analysis of DNA from samples of body tissues or fluids in order to identify individuals. Fragments of the DNA are amplified using PCR (The polymerase chain reaction (PCR) is a technique used in molecular biology to amplify a single copy or a few copies of a piece of DNA across several orders of magnitude, generating thousands to millions of copies of a particular DNA sequence) and the lengths of the resulting DNA pieces are characteristic of each individual. Because of the PCR step, only very small amounts of DNA are needed for DNA profiling.

DNA polymerase: The DNA polymerases are enzymes that create DNA molecules by assembling nucleotides These enzymes are essential to DNA replication and usually work in pairs to create two identical DNA strands from one original DNA molecule. There are several DNA polymerases in our cells; and although they have slightly different functions, they all are involved in DNA replication.

DNA replication: n molecular biology, DNA replication is the biological process of producing two identical replicas of DNA from one original DNA molecule. This biological process occurs in all living organisms and is the origin for biological inheritance.

Dominant: Refers to the allele that is expressed when two different alleles are found together. Dominance in genetics is a relationship between alleles of one gene, in which the effect on phenotype of one allele conceals the contribution of a second allele at the same locus. The first allele is dominant and the second one is recessive.

Embryo: An embryo develops from a zygote, the single cell resulting from the fertilization of the female egg by the male sperm cell. It develops into a free-living miniature adult or larva, in animals, or germinates into a seedling, in plants.

Enucleation: Enucleation is the removal of a nucleus from its cell. This process is used in the cloning of organisms or creatures.

Enzyme: A protein that can speed up chemical reactions without getting chemically changed itself. Enzymes are biological molecules that act as catalysts and help complex reactions occur everywhere in life. For example, Human saliva contains an enzyme called amylase that speeds up the chemical reaction of converting starches in our food to sugars.

Epidemiology: The field of medicine that deals with the incidence, distribution, and potential control of diseases and other factors relating to health.

Epigenetics: The study of changes in organisms caused by modification of gene expression rather than alteration of the genetic code itself. Such modifications include DNA, heterochromatic formation, genomic imprinting and X-chromosome inactivation.

Eukaryote: A eukaryote is an organism with multiple cells, or a single cell with a complicated structure. In these cells the genetic material is organized into chromosomes in the cell nucleus. Animals, plants, algae and fungi are all eukaryotes.

Exon: The region of RNA that is translated into protein. The term exon refers to both the DNA sequence within a gene and to the corresponding sequence in RNA transcripts. In eukaryotes, the primary RNA usually contains several exons that are separated by introns. Before the RNA is translated, the introns are cut off and only exons remain.

Gamete: A mature haploid male or female germ cell that is able to unite with another of the opposite sex in sexual reproduction to develop a zygote.

Gel electrophoresis: It is a technique used in laboratories to split mixtures of DNA, RNA, or proteins according to molecular size.

Gene: A gene is the basic physical and functional unit of heredity. Genes, which are made up of DNA, encode messages for the synthesis of proteins and functional RNAs. Genes help determine an organism's appearance and behavior.

Gene editing: It is a type of genetic engineering in which DNA is inserted, deleted or replaced in the genome of an organism using engineered nucleases, or molecular scissors. The three common types of gene editing methods are Talen, Zinc finger and Crispr.

Gene expression: Gene expression is the process by which information from a gene is used in the synthesis of a functional gene product. These products are often proteins, but in non-protein coding, the product is a functional RNA. Either way, the gene's information is out there to impact the individual' phenotype.

Gene knockout: It is an existing gene by replacing it or disrupting it with an artificial piece of DNA. Knockout mice are often used to study the function of the knocked-out gene by noting what differences are observed in the mutated mouse relative to a normal mouse.

Gene therapy: T he transplantation of normal genes into cells in place of missing or defective ones in order to correct genetic illness. In the future, this technique may allow doctors to treat a disorder by inserting a gene into a patient's cells instead of using drugs or surgery.

Genetic engineering: The deliberate modification of the characteristics of an organism by manipulating its genetic material. Many different techniques are used in genetic engineering, including the examples of gene and trangenic organisms.

Genetic Marker: A segment of DNA that can be tracked from one generation to the next. A gene or short sequence of DNA used to identify a chromosome or to locate other genes on a genetic map. Markers can be entire genes or a single letter of code.

Genetic testing: Using tests to diagnose or determine the predisposition to a genetic disease, The sequencing of human DNA in order to discover genetic differences, anomalies, or mutations that may prove pathological. The type of test can vary and includes tests directly on DNA, as well as biochemical tests that analyze proteins or metabolites linked to genetic diseases. The tests can also be used to prove paternity.

Genetically Modified Organism (GMO): When a gene from one organism is purposely moved to improve or change another organism in a laboratory, the result is a genetically modified organism. It is also sometimes called "transgenic" for transfer of genes. There are different ways of moving genes to produce desirable traits. When scientists generate GMOs, they combine existing pieces of DNA in new ways to give organism new characteristics.

Genome: A genome is an organism's complete set of DNA – basically a blueprint for an organism's structure and function. In other word, it is the complete set of genes or genetic material present in a cell or organism.

Genomics: Genomics is the science that aims to decipher and understand the entire genetic information of an organism (ie: plants, animals, humans, viruses and microorganisms) encoded in DNA and corresponding complements such as RNA, proteins and metabolites. Experts in genomics strive to determine complete DNA sequences and perform genetic mapping to help understand disease.

Genomic imprinting: When the expression of the maternally derived or the paternally derived allele of a gene is suppressed in the embryo. If the allele inherited from the father is imprinted, it is thereby silenced, and only the allele from the mother is expressed. Gene inactivation is correlated with increased DNA methylation of the gene.

Genotype: The genotype is the part (DNA sequence) of the genetic makeup of a cell, and therefore of an organism or individual. The specific set of alleles contained in the DNA of an organism. The genotype, as well as environmental and epigenetic factors, determines the final traits.

Germline: The group or line of cells that gives rise to reproductive cells (sperm or eggs). A series of germ cells each descended from earlier cells in the series, regarded as continuing through successive generations of an organism. Mutations in the germline are passed on to future generations. Cells that are not part of the germline are called somatic cells.

Haploid: Haploid describes a cell that contains a single set of chromosomes and therefore only one allele of each gene. Human sperm and egg cells, for example, are haploid. When they combine during fertilization, they form a diploid.

Heritability: Heritability is defined as the degree to which individual genetic variation accounts for phenotypic variation seen in a population. If a trait has a high heritability it generally means that genetic factors strongly influence the amount of variation.

Heterozygous: A diploid organism is heterozygous at a gene locus when its cells contain two different alleles of a gene.

Histone: Histones are proteins found in eukaryotic cell nuclei that package and order the DNA into structural units called nucleosomes. A group of structural proteins that act as spools around which DNA winds.

Homologous: Having similar structure and anatomical position (but not necessarily the same function) in different organisms suggesting a common ancestry or evolutionary origin Homology can be subdivided into orthology and parology.

Homozygous: When an individual has two of the same allele, whether dominant or recessive, they are homozygous. Heterozygous denotes having one each of two different alleles. Recessive traits only appear if an individual is homozygous for the corresponding gene..

Human genome project: The Human Genome Project (HGP) is an international scientific research project with the goal of establishing the sequence of chemical base pairs which make up human DNA, and of identifying and mapping all of the genes of the human genome from both a physical and a functional standpoint. The reason behind the project was that sequencing and identifying all human genes would help us to better understand the genetic roots of disease and find ways to diagnose, treat and perhaps prevent many diseases.

In vitro: Biological processes and reactions occurring in either (i) cells or tissues grown in culture or (ii) cell extracts or synthetic mixtures of cells components. In general, In vitro fertilization (or fertilization; IVF) is a process by which an egg is fertilized by sperm outside the body

In vivo: A process performed in a living organism.

Inherited: Heredity is the genetic information passing for traits from parents to their offspring, either through asexual reproduction or sexual reproduction. The genes present in the parents are passed down to their offspring through egg or sperm cells.

Intron: Introns are noncoding sections of an RNA transcript, or the DNA encoding it, that are spliced out before the RNA molecule is translated into a protein. Not all parts of RNA leave the nucleus and introns are the fragments that are cut out.

Junk DNA: It is genomic DNA that does not encode proteins, and whose function, if it has one, is not well understood. These DNA segments were thought to be useless.

Ligase: An enzyme that can join pieces of DNA or another substance.

Meiosis: Meiosis is a process where a single cell divides twice to produce four cells containing half the original amount of genetic information. These cells are sex cells – sperm in males, eggs in females. Meiosis is the mechanism by which gametes are produced.

Mendel, Gregor: Gregor Mendel was an Austrian monk who discovered the basic principles of heredity through experiments in his garden. He provided a collection of experimental observations that were translated into generally applicable rules describing how some traits are passed on between generations.

Mendelian inheritance: A set of principles of inheritance derived from the work of Gregor Mendel. In short, these principles state that alleles separate into gamets such that each gamete contains only a single copy of a gene (segregation). Furthermore, the alleles of different genes separate into gametes independently and do not sort based on the inheritance of other genes (independent assortment).

Messenger RNA (mRNA): An RNA molecule which carries the message that acts as a template for translation into protein. Messenger RNA (mRNA) conveys genetic information from DNA to the ribosome, where they specify the amino acid sequence of the protein products of gene expression.

Microbe: Microbes are single-cell organisms so tiny that millions can fit into the eye of a needle. They are the oldest form of life on earth. A microbe is a microscopic organism; a micro-organism. Include bacteria, some fungi, protozoa and viruses.

Microorganism: Microorganisms are microscopic, living, single-celled organisms such as bacteria or virus.

Microtubules: Microtubules are found throughout the cytoplasm They are fine hollow protein tubes involved in the intracellular transport of materials and movement of organelles. All of the microfilaments and microtubules combine to form the cytoskeleton of the cell. Microtubules also form the mitotic and meiotic spindles.

Mitochondria: An organelle found in large numbers in most cells, in which the biochemical processes of respiration and energy production occur An organelle found in the cytoplasm of most eukaryotic cells and responsible for generating the energy for the cell.

Mitochondrial DNA (mtDNA): Mitochondrial DNA (mtDNA or mDNA) is the DNA located in mitochondria, cellular organelles within eukaryotic cells that convert chemical energy from food into a form that cells can use, adenosine triphosphate (ATP). mtDNA is a circular DNA molecule found inside each eukaryotic mitochondrion.

Mitosis: Mitosis is a part of the cell cycle in which chromosomes in a cell nucleus are separated into two identical sets of chromosomes, and each set ends up in its own nucleus. Cell division that produces two daughter cells with Chromosomes, and therefore DNA that is identical to the parent cell. Mitosis is the mechanism by which somatic cells are produced.

Multifactorial: Refers to a biological or physiological observation that is attributable to many factors. For example, some traits are determined by the interplay between genes and environment. Most common traits like skin color and height are multifactorial. Although your genes give a rough estimate of your height, environment can also impact your height because if you don't have adequate nutrition during childhood, you will likely not reach your height potential.

Mutation: A change in the DNA sequence of an organism that can have no effect, or be either beneficial or harmful.

Necrosis: Localized and premature death of cells in an organ or tissue due to disease or injury.

Nucleic Acid: A large organic molecule made up of a chain of nucleotide. Examples include DNA and RNA.

Nucleoside: Organic compound made up of a purine or pyrimidine (base) joined to a sugar. Nucleosides are structurally very similar to nucleotide but, unlike them, they do not contain any phosphate groups.

Nucleosome: A basic unit of DNA packaging in eukaryotes, consisting of a segment of DNA wound around eight histone protein cores. This structure is often compared to thread wrapped around a spool.

Nucleotide: Organic compound made up of a purine or pyrimidine (base) joined to a sugar and a phosphate group. Nucleic acids (DNA & RNA) contain nucleotides linked together in long chains.

Nucleus: A membrane-bound organelle containing the genetic material of a eukaryotic cell.

Nutrigenomics: The study of the interaction between diet, genes and environment and how they affect human health.

Oligonucleotide: A short fragment of single-stranded DNA or RNA that is typically 5-50 nucleotides long. Oligonucleotides can be primers to start PCR and also the fixed target in DNA microarrays.

Oncogene: A gene that, when mutated, can promote growth beyond the cell's normal needs, thus leading to tumours. For example, a growth factor gene that is always expressed, even when there is no signal for growth, can cause a cell to grow beyond its normal limits.

Operon: A segment of DNA containing linked genes that function in a coordinated manner, usually under the control of a single promoter.

Organism: An individual animal, plant, or single-celled life form.

Orthologous: Equivalent genes in different species that are homologous because they have both evolved in a direct line from a common ancestral gene.

PAM: Protospacer-Adjacent Motif

Parasite: An organism that lives in or on another organism (its host) and benefits by deriving nutrients at the host's expense.

Pathogen: An agent causing disease.

Pathology: The study of diseases and their causes, processes, development and consequences.

Paralogous: Two homologous genes in the same or different genomes that are similar because they derive from a gene duplication.

Penetrance: The proportion of individuals carrying a particular genotype that express the associated phenotype. Important for genetic diseases because complete penetrance means that 100% of people with the disease mutation will have the associated phenotype.

Peptide bond: A covalent bond joining the alpha-amino group of one amino acid to the carboxyl group of another with the loss of a water molecule. It is the bond linking amino acids together in a protein chain.

Phagocytosis: The process of a cell engulfing and eating particles or other cells.

Pharmacogenomics: A branch of pharmacology concerned with using DNA and amino acid sequence data to inform drug development and testing. An important application of pharmacogenomics is correlating individual genetic variation with drug responses.

Phenotype: The set of observable characteristics of an organism that are the result of its genotype and the environment.

Plasmid: The small circular genetic material present in bacterial cells and used in genetic engineering or genetic modification.

Polygenic: Polygenic traits are affected by more than one gene. phenotype variation in some traits is due to the interaction of many genes, each with a small additive effect on the character in question. For example, it is currently believed that there are at least 11 genes involved in the determination of skin color.

Polymerase Chain Reaction (PCR): A technique for selectively and rapidly replicating a particular stretch of DNA in vitro to produce a large amount of that particular sequence.

Preimplantation Genetic Diagnosis (PGD): A technology that allows embryos created by in vitro fertilization to be tested for a genetic condition before transferring them to a uterus. PGD is an option for couples who are at risk of passing on a genetic condition. Once the embryo has grown to the 8-16 cell stage, one cell is removed and genetic testing is done. Only those embryos that are not affected with the genetic condition tested for are implanted.

Primer: A short nuceic acid chain that serves as a starting point for the copying of DNA (DNA replication). This short stretch of DNA or RNA is complementary to part of the DNA that is about to be copied, and binds to it to allow attachment of the other machinery needed (e.g. DNA polymeras) to copy the DNA.

Prokaryote: An organism that consists of cells which do not have membrane-bound organelles. Most prokaryotes are single-celled organisms. Importantly, the DNA of prokaryotes is found loose in the cell rather than in a nucleus.

Promoter: A region of DNA located at the beginning (5' end) of a gene. It contains sequences important in starting transcription. Mutations in the promoter region may cause incorrect expression of the gene and can lead to disease.

Protein: Large organic molecule made up of various combinations of amino acids. Proteins support living organisms' shape and structure; carry messages within cells and between them; and as enzymes they regulate the chemical processes that sustain life.

Protein Synthesis: The process of building proteins by reading messages carried on mRNA and converting them to chains of amino acids using ribosomes and transfer RNA (tRNA).

Proteomics: The science that studies which proteins of the genome are expressed and when. Initially aimed at cataloguing the proteins present in a cell under various conditions, proteomics has joined up with genomics to try to understand how the expression of the genome enables all the complex functions of the cell to work.

Recessive: A heritable characteristic controlled by genes that is expressed in offspring only when inherited from both parents (ie. Homozygous recessive).

Recombination: Reorganization, shuffling or other moving of genetic material from one place on the chromosome to another, or to a different chromosome. The most common example is crossing over.

Restriction enzyme: An enzyme that can cut DNA in specific places in the DNA molecule.

Restriction Fragment Length Polymorphism (RFLP): A size difference (length polymorphism) among fragments after DNA has been cut with restriction enzymes. Each cut piece is a restriction fragment and each fragment size is affected by DNA sequence. For example, if a target sequence is present, the restriction enzyme will cut the DNA molecule; if a target sequence is missing, the DNA won't be cut and the fragment will consequently be longer. This characteristic has been important in creating lab tests for paternity and for DNA fingerprinting.

Ribonucleic Acid (RNA): A long nucleic acid molecule found in the nucleus and cytoplasm of a cell. Similar to DNA, RNA consists a sugar-phosphate backbone with nitrogenous bases. RNA differs from DNA by having a different sugar in its backbone (ribose instead of deoxyribose); having uracil as a base instead of thymine; and functioning as a single-stranded molecule instead of a double-stranded helix. One function of RNA is to convey genetic information, encoded by DNA, to the protein synthesis machinery. This process is known as translation and involves three types of RNA that work together to achieve this task: mRNA, rRNA and tRNA.

gRNA: guide RNA

sgRNA: single guide RNA

tracrRNA, trRNA**:** trans-activating crRNA

Ribosomal RNA (rRNA): The RNA component of the ribosome. It works together with mRNA, tRNA and amino acids during translation.

Ribosome: A complex of proteins and rRNA that converts information from mRNA into an amino acid chain. In other words: it translates the mRNA sequence into a protein.

RNA polymerase: The enzyme which transcribes DNA into RNA.

Semidominant: Having an intermediate phenotype in an individual that has a heterozygous genotype for a particular trait. For example, degree of hair curliness is semidominant: if "A" represents curly hair, and "a" represents straight hair, an individual who has an "Aa" genotype would have wavy (i.e. not curly but not straight, rather intermediate) hair.

Sequencing: Reading of the components of a molecular chain of building blocks. For example, determining the order of nucleotides in a DNA or RNA chain or the amino acids within a protein.

Sex-linked traits: The tendency of certain characteristics to appear in one sex. traits encoded by genes on one of the sex chromosomes (X or Y chromosomes in humans) can be expressed differently in males and females because males have an X and a Y chromosome, whereas females have two X chromosomes.

Side chain: In an amino acid, that part not involved in forming the peptide bond. The side chain gives each amino acid its characteristic chemical and physical properties.

Single Nucleotide Polymorphism (SNP): A variation in a single base (A, T, C or G) within a sequence of DNA. For any single base variation to be called a SNP the minor allele must be found in more than 1% of the population. So far more than 6 million SNPs have been discovered in the human genome. SNPs do not generally cause disease directly but some SNPs may indicate an individual's susceptibility to disease or the response to drugs and other treatments.

Somatic Cell Nuclear Transfer (SCNT): A laboratory method for creating clones of animals. SCNT requires 2 donors: a nucleus and an egg. The egg is stripped of its nucleus and the donor nucleus is placed inside. Because the donor nucleus is diploid, whereas normal egg cells are haploid, the egg is "tricked" into behaving as if it were fertilized. This leads the egg to divide and differentiate, resulting in an embryo that will develop into a copy (clone) of the organism that donated the nucleus.

Somatic Cells: The cells that form the body only. Unlike in germ cells, mutations or manipulations in these cells are not passed on to the next generation.

Spindle: The structure formed by microtubules stretching between opposite sides of the cell during mitosis or meiosis and which guides the movement of chromosomes.

Sporadic: Rare or unpredictable. In genetics, specifically means that it is not known to be inherited. For example, if one person in a family had a genetic condition that was not found in any other family members, they may be referred to as having a sporadic case of that condition.

Stem Cell: A cell that has the ability to self-renew or divide indefinitely. A cell found in foetuses, embryos, and some adult tissues that can give rise to a wide range of other cells. Stem cells can also develop into other specialized cell types. There are several classifications: totipotent, pluripotent, and multipotent.

TALEN: Transcription-Activator Like Effector Nuclease

Telomere: A region of repetitive DNA at the end of a linear chromosome that serves to protect the end from deterioration or destruction. In somatic, with each round of cell division, telomeres get shorter and once they become too short to protect the ends of the chromosomes, cell division may be prevented, or cells may die.

Teratogen: A substance or exposure that causes birth defects. Many drugs, infections, chemicals and radiation can be teratogens.

Trait: A characteristic that an individual possesses. It could be a personality or behavior trait (like being warm and friendly), intelligence (a high IQ score) or a physical feature (like red hair or height). Genes may or may not have a determining role in a trait.

Transcript: The RNA that is synthesized by RNA polymerase on a DNA or RNA template.

Transcription: The enzymatic process where complementary RNA sequences are created from DNA templates.

Transcriptomics: The complete set of RNA transcripts produced by the genome at any given time.

Transfer RNA (tRNA): RNA responsible for bringing amino acids to the ribosome and working with the mRNA and rRNA to make an amino acid chain. tRNA transfers the amino acid to the growing protein chain during translation. Each tRNA molecule is specific to a certain amino acid.

Transgenic: Having a piece of DNA that originally stems from a different species. Transgenic organisms are a subset of Genetically Modified Organisms (GMOs).

Translation: The process of producing proteins from the information stored on an mRNA molecule. In other words it is the process of translating from the language of nucleotides to the language of proteins.

Translocation: The transfer of part of a chromosome onto another nonhomologous chromosome. Translocations may be balanced (having the right amount of chromosome material) or unbalanced (having too much or too little chromosome material). A reciprocal translocation is where two nonhomologous chromosomes exchange material.

Transposon: A segment of DNA that can remove itself from a chromosome and insert itself somewhere else in the genome, i.e. it can 'jump' from one place in the genome to another. Transposons can also move between organisms. They play a role in the transfer of antibiotic resistance among bacteria and can cause disease by creating mutations.

Triplet repeat expansion: Triplet repeats are 3 nucleotides that are repeated a number of times (e.g., TACTACTACTACTAC). Triplet repeat expansion is when extra repeats are added (e.g., more TACs). If this large repeat is within a gene, it can cause disease (e.g. Huntington's disease).

Trisomy: The occurrence of an extra chromosome (ie: third copy of a particular chromosome) in the total chromosome count of an individual. People typically have two copies (disomy) of all of their chromosomes. Often, having trisomy of a full chromosome is not compatible with life except for an additional chromosome 21 (Down Syndrome) or extra sex chromosomes.

Trisomy 18 and Trisomy 13: Typically lethal conditions caused by an extra chromosome in each cell (chromosomes 18 and 13 respectively) and involving many birth defects.

Trisomy 21: Also known as Down syndrome. Caused by an extra chromosome 21 in each cell.

Tumor: A growth resulting from the abnormal proliferation of cells. It may be self-limiting or non-invasive, when it is called a benign tumor. A malignant tumor or cancer is when a tumor proliferates indefinitely and invades underlying tissues and metastasizes.

Tumor Suppressor Gene: A normal gene that controls how often and how fast a cell divides. If both copies of a tumor suppressor gene are inactivated, the growth of the cell may go out of control and become a tumor.

Uterus: In female mammals, the organ in which the embryo develops and is nourished before birth. In other mammals, the uterus is an enlarged portion of the oviduct modified to serve as a place for development of young or of eggs.

Variable-Number Tandem Repeat (VNTR): Catch-all term for repeats in the genome. Generally not associated with disease, these repeats are used in DNA profiling because the sequences vary in length between people. VNTR sounds complicated, but the phrase simply means, "repeats that vary in size".

Vector: The way in which genetic material is transferred from a donor to a recipient, e.g. viruses, bacterial cells or plasmids.

Virus: An intracellular obligate parasite that is unable to multiply or express its genes outside a host cell as it requires host cell enzymes to aid DNA, transcription and translation. Viruses cause many diseases of man, animals, plants and bacteria.

X chromosome: One of the two sex chromosomes in humans. Women generally have two X chromosomes whereas men have one X chromosome and one Y chromosome.

X chromosome inactivation: The inactivation of all but one copy of the X chromosome in female mammals.

Y chromosome: One of the two sex chromosomes in humans. Men generally have one X chromosome and one Y chromosome whereas women have two X chromosomes. The Y chromosome contains the genes that trigger male development and proper sperm formation.

ZFN: Zinc-Finger Nuclease

Zygote: A diploid cell that is the result of the fusion of a haploid egg and haploid sperm. The term zygote only applies before cell division starts (after which it is called an embryo).

Printed in the United States
By Bookmasters